Learning Materials in Biosciences

Learning Materials in Biosciences textbooks compactly and concisely discuss a specific biological, biomedical, biochemical, bioengineering or cell biologic topic. The textbooks in this series are based on lectures for upper-level undergraduates, master's and graduate students, presented and written by authoritative figures in the field at leading universities around the globe.

The titles are organized to guide the reader to a deeper understanding of the concepts covered.

Each textbook provides readers with fundamental insights into the subject and prepares them to independently pursue further thinking and research on the topic. Colored figures, step-by-step protocols and take-home messages offer an accessible approach to learning and understanding.

In addition to being designed to benefit students, Learning Materials textbooks represent a valuable tool for lecturers and teachers, helping them to prepare their own respective coursework.

More information about this series at http://www.springer.com/series/15430

Cesare Terracciano
Samuel Guymer

Editors

Heart of the Matter

Key concepts in cardiovascular science

 Springer

Editors
Cesare Terracciano
National Heart & Lung Institute
Imperial College London
London, United Kingdom

Samuel Guymer
Imperial College London & St George's
Medical School
London, United Kingdom

ISSN 2509-6125   ISSN 2509-6133 (electronic)
Learning Materials in Biosciences
ISBN 978-3-030-24218-3   ISBN 978-3-030-24219-0 (eBook)
https://doi.org/10.1007/978-3-030-24219-0

This Springer imprint is published by the registered company Springer Nature Switzerland AG
The registered company address is: Gewerbestrasse 11, 6330 Cham, Switzerland

"Fatti non foste a viver come bruti,

ma per servir virtute e conoscenza"

Dante Alighieri, La Divina Commedia, Inferno, Canto XXVI, vv119–120

Preface by Cesare Terracciano

"Not another textbook!" I hear you saying! In the era of digital technology, powerful web browsers, instantaneous globalisation and online lectures, why do we need a textbook in cardiovascular sciences? Knowledge, defined as the accumulation of pure information, is no longer the prerogative and currency of university education, and anyone with a decent smartphone can access all the information available from every part of the world at the press of a few buttons. In the past, even until quite recently, textbooks had a central role in education; they were memorised almost word for word, but now that this information is promptly available using other media and is comprehensive and up-to-date, why do we need a textbook in the learning process of medical sciences?

The BSc course in Cardiovascular Sciences has been running for approximately 20 years and is the main undergraduate cardiac teaching activity of the National Heart and Lung Institute; the course was established when the institute joined Imperial College London in the mid-1990s. The course has changed substantially over the years, following the massive changes in university learning, associated with the exponential growth in scientific discovery, the diffusion of digital technologies and the Internet. Starting as a module to complement both the BSc in Biomedical Sciences and the Science year of the MBBS curriculum, the course was originally based almost entirely on the basic biology of the cardiovascular system. Over the years, becoming more integrated in the teaching of Medicine, the pathophysiology aspects have become more important, and cardiac diseases are now the main study subjects of the course. Far from being purely clinical, the main objective of the course is to explore the scientific foundation of cardiac disease, and, in the process, it offers the opportunity to learn about biology, physiology and regulation of the cardiovascular system. For this reason, the course is very popular with medical students who thoroughly enjoy the learning of science within the context of clinical problems.

The course is taught by a faculty consisting almost entirely of academics whose primary role is to perform world-leading cardiovascular research. Many of our teachers offer the invaluable experience of being directly involved in the scientific discoveries they teach. Their enthusiasm and understanding of how disease happens, why certain people are affected and others are protected, why a particular treatment should be used and how do we know that it works are passed onto the students together with an extraordinary level of expertise. This is the main factor that has made our course very popular over the years.

In this context, the value of university learning shifts from bare memorising of information to its critical analysis. What the students should gain are guidance, opinions, interpretations and experience. Active learning, where students are engaged directly and challenged, where information is dissected and analysed in sessions led by both the student and the lecturer, where controversies, practical and hands-on applications are the main focus, is the way forward for our course and university learning in general.

Active learning activities are only rewarding if the students have already some knowledge of the topic, ask more pertinent questions and make more appropriate comments. The problem with this approach is in the lack of fundamental knowledge from the students during the lecture, one that, in the traditional way of thinking, they have come to acquire. So a vicious circle ensues, where students need to memorise basic information before they can be engaged in active learning. They will spend the best part of the course learning facts that do not require interaction with teachers and only at the end, often after the assessments, are they sufficiently ready for active learning. We must ensure that the students acquire the necessary knowledge *before* they engage with the lecturers. This information must be manageable, vetted, up-to-date and sufficient for engaging in intelligent and critical analysis of the subject, to allow further and in-depth discussions and carry out practical sessions. This textbook is designed to fulfil this need: the book contains the starting point, the necessary knowledge to engage in the course. In addition, the book has also a crucial role in allowing students coming from different streams, colleges or different education systems to approach the course from a similar starting point, helping every student to engage and thrive.

In order to fulfil this background role, this textbook has two main features: firstly, it is simple and broad in its scope. It only requires a basic understanding that every student at this stage of learning should have. It is not designed to be comprehensive but to provide a manageable list of reading material that should kick-start and engage the students to proceed in the correct direction when they need more in-depth details. Secondly, the chapters have been written by our teachers in close collaboration with previous students of the BSc course in Cardiovascular Sciences. Each chapter is authored by at least one student and one lecturer. The contribution of the lecturer is to ensure scientific rigour, topical and relevant directions and experience, while the students contribute by suggesting the correct level of pitching, a more amenable language and a modern twist. The combination of these two roles results in an essential learning tool, a springboard to be used by everyone around the world with a biomedical background who wants to start the journey into science in the context of cardiac disease.

Cesare Terracciano
London, UK
23 January 2019

Preface by Samuel Guymer

The heart represents an intellectual endeavour like no other: an apex of contradiction where innumerable interacting mechanisms of increasing complexity converge to create a simple pumping apparatus that propels the volume of a double espresso into our vasculature each minute, the subject of innumerable research but equally the repository of the soul. As a student of medicine, it was this contradiction that gravitated my curiosity toward this organ like so many before me. How can we know so much about the heart yet still have so many questions? We are, in the words of the Persian poet Ibn Yamin, "the man who doesn't know, but knows that he doesn't know."

The Cardiovascular Sciences BSc programme was the catalyst for my realisation (and acceptance) of this identity: a course that provides students with the tools and knowledge to challenge the consensus, to question *why* a particular guideline recommends a specific investigation or therapy. It is a programme that assumes its responsibility to nourish the intellectually hungry whilst simultaneously immersing them in the controversies and unknowns that persist in this field. An approach that stands in stark contrast to a medical education that so often retreats into a complacency of lionising the memorisation of things we *know*, e.g. that drug x at dose y is indicated in z condition.

This shift in the learning paradigm is exemplified by the contrasting embryological theories of sinoatrial nodal origin, both of which are assiduously dissected down to a molecular level of detail with no definitive answer as to which is "the truth." Instead, students are provided the respective evidence bases and encouraged to find "the truth" for themselves: Is it one of the two predominant theories, some interspersion of both, or perhaps neither? We have attempted to encapsulate this learning mentality within this book, with most chapters containing a section entitled "What We Don't Know". It is here that an area lacking a definitive "truth" is expanded upon, allowing you to critically appraise the evidence base(s) and reach a conclusion for yourself.

Another salient feature of this textbook is the synergy of student and lecturer to create material that covers the fundamentals, the "heart of the matter", whilst concurrently exploring new developments and research. In this way, we are able to fully utilise the pioneering work being undertaken at the National Heart and Lung Institute.

Ultimately, the aim of this textbook is to equip you with an understanding of cardiovascular sciences and where the deficits in understanding persist, allowing the finitude of your knowledge to be challenged by the infinitude of your curiosity and pursuit for greater understanding of this contradiction of an organ.

Samuel Guymer
London, UK
8 February 2019

Acknowledgements

We would like to thank Mr. Barrett Downing for the invaluable editorial support at all stages of the preparation of this textbook. We are grateful to our illustrator, Ms. Umamah Tarvala, for providing many excellent figures and graphs. Finally, we would like to thank all the past and present students of the BSc course in Cardiovascular Science at Imperial College London; their feedback has been essential for the conceptualisation of this book.

Contents

Contributors

Ayesha Ahmed
Imperial College London
London, UK

The University of Sheffield
Sheffield, UK
aahmed15@sheffield.ac.uk

Ahran D. Arnold
National Heart and Lung Institute
Imperial College London
London, UK
ahran.arnold@imperial.ac.uk

Yasmin Bashir
Imperial College London
London, UK
yasmin.bashir12@imperial.ac.uk

Graeme Birdsey
National Heart and Lung Institute
Vascular Sciences, Imperial College London
London, UK
g.birdsey@imperial.ac.uk

Joseph J. Boyle
National Heart and Lung Institute
Vascular Sciences, Imperial College London
London, UK
joseph.boyle@imperial.ac.uk

Thomas Brand
National Heart and Lung Institute
Myocardial Function, Imperial College London
London, UK
t.brand@imperial.ac.uk

Giles Chick
Imperial College London
London, UK
giles.chick14@imperial.ac.uk

Rasheda Chowdhury
National Heart and Lung Institute
Myocardial Function, Imperial College London
London, UK
r.chowdhury@imperial.ac.uk

Liam Couch
National Heart and Lung Institute
Myocardial Function, Imperial College London
London, UK
liam.couch11@imperial.ac.uk

James Crawley
Centre for Haematology,
Imperial College London
London, UK
j.crawley@imperial.ac.uk

Ben N. Cullen
Imperial College London
London, UK
ben.cullen14@imperial.ac.uk

Pieter P. de Tombe
Department of Physiology and Biophysics
College of Medicine, University of Illinois
Chicago, IL, USA
pdetombe@uic.edu

Zarius Ferozepurwalla
Imperial College London
London, UK
zarius.ferozepurwalla14@imperial.ac.uk

Jack Griffiths
Imperial College London
London, UK
jack.griffiths14@imperial.ac.uk

Samuel Guymer
St George's, University of London
London, UK

National Heart and Lung Institute
Myocardial Function, Imperial College London
London, UK
samuel.guymer17@imperial.ac.uk

Dominic Gyimah
Brighton and Sussex Medical School
Brighton, UK

Imperial College London
London, UK
d.gyimah1@uni.bsms.ac.uk

Balvinder Handa
National Heart and Lung Institute
Myocardial Function, Imperial College London
London, UK
balvinder.handa05@imperial.ac.uk

Moghees Hanif
Queen Mary University of London
London, UK

Imperial College London
London, UK
moghees_hanif@me.com

Angeliki Iakovou
Queen Mary University of London
London, UK

Zohya Khalique
National Heart and Lung Institute,
Imperial College London
London, UK
z.khalique@imperial.ac.uk

Praveena Krishnakumar
Imperial College London
London, UK
praveena.krishnakumar14@imperial.ac.uk

Kenneth T. MacLeod
National Heart and Lung Institute
Imperial College London
London, UK
k.t.macleod@imperial.ac.uk

Justin Mason
National Heart and Lung Institute
Vascular Sciences, Imperial College London
London, UK
justin.mason@imperial.ac.uk

Kabir Matwala
Imperial College London
London, UK
kabir.matwala14@imperial.ac.uk

Jude Merzah
Imperial College London
London, UK

Queen's University Belfast
Belfast, UK
jude.merzah@hotmail.com

Salomon Narodden
National Heart and Lung Institute
Imperial College London
London, UK
s.narodden@imperial.ac.uk

Fu Siong Ng
National Heart and Lung Institute
Myocardial Function, Imperial College London
London, UK
f.ng@imperial.ac.uk

Fotios G. Pitoulis
National Heart and Lung Institute
Myocardial Function, Imperial College London
London, UK
fotios.pitoulis14@imperial.ac.uk

Haaris Rahim
Imperial College London
London, UK
haaris.rahim14@imperial.ac.uk

Mridul Rana
Imperial College London
London, UK
mridul.rana14@imperial.ac.uk

Rohin K. Reddy
Hull York Medical School
York, UK

Imperial College London
London, UK

Alec Saunders
Brighton and Sussex Medical School
Brighton, UK

National Heart and Lung Institute,
Myocardial Function, Imperial College London
London, UK
bsms5386@uni.bsms.ac.uk

Michael Schachter
National Heart and Lung Institute
Myocardial Function, Imperial College London
London, UK
m.schachter@imperial.ac.uk

Mayooran Shanmuganathan
National Heart and Lung Institute
Myocardial Function, Imperial College London
London, UK

Cardiovascular Medicine, University of Oxford
Oxford, UK
mayooran.shan@lmh.ox.ac.uk

Shiv-Raj Sharma
Imperial College London
London, UK
shiv-raj.sharma14@imperial.ac.uk

Umamah Tarvala
Barts and The London School of Medicine
and Denstistry (QMUL)
London, UK

Beth Taylor
Imperial College London
London, UK
bethany.taylor14@imperial.ac.uk

Cesare M. Terracciano
National Heart and Lung Institute
Myocardial Function, Imperial College London
London, UK
c.terracciano@imperial.ac.uk

Lieze Thielemans
Imperial College London
London, UK
lieze.thielemans14@imperial.ac.uk

Elizabeth Thong
Imperial College London
London, UK
elizabeth.thong14@imperial.ac.uk

Adam Tsao
Imperial College London
London, UK
adam.tsao14@imperial.ac.uk

Zachary I. Whinnett
National Heart and Lung Institute
Imperial College London
London, UK
z.whinnett@imperial.ac.uk

Anatomy of the Heart and Coronary Vasculature

Dominic Gyimah, Alec Saunders, Elizabeth Thong, and Balvinder Handa

© Springer Nature Switzerland AG 2019
C. Terracciano, S. Guymer (eds.), *Heart of the Matter*, Learning Materials in Biosciences,
https://doi.org/10.1007/978-3-030-24219-0_1

What You Will Learn in This Chapter

This chapter will explore the anatomy of the heart and its vasculature, discussing key structures and their functional role within the organ's contractility. A solid anatomical understanding will aid the learning of this complex organ's role in the body and the consequences of morphological abnormalities such as malformations. Finally, the potential role of three-dimensional (3D) printing will be discussed as a novel technique of visualising anatomical structures.

Learning Objectives
- Analyse the gross anatomy of the heart, including the four chambers, valves, great vessels and coronary vasculature.
- Appreciate the role of the underlying principles behind imaging techniques such as cine MRI and late gadolinium enhancement.
- Begin to think about the potential role of neoteric modelling techniques, namely 3D printing.

1.1 The Evolution of our Understanding

During the evolution of vertebrates, the cardiovascular system has undergone significant anatomical and functional changes. A contractile vessel characterised the heart of the earliest vertebrates, with peristaltic movements providing the perfusion of the vasculature at low pressures. Provided organisms were small, simple diffusion sufficed. However, as living beings grew in size, a more sophisticated system was required to transport oxygen, nutrients and remove waste products. A heart with a single ventricle and atrium evolved later among fish. Later, birds and mammals evolved a four-chambered heart with a septum between the aortic and pulmonary valves to allow for segregated circulations. Accompanying this division between the systemic and pulmonary circulations was an increase in heart rate with the aim of augmenting cardiac output to meet the requirements of a higher basal metabolic rate [1].

The nature, function and anatomy of the heart as we know it today is due to the culmination of extensive study dating back to 3500 BC [2]. In its youth, the main focus of anatomical studies was to determine the origin of intelligence, soul and mind. However, the Hippocratic and post-Hippocratic era marked a shift in paradigm, whereby findings became more 'science-like' as supported by dissections. Hippocrates described the heart as having two ventricles connected by orifices through the interventricular septum. This morphological difference was attributed to the left ventricle being the site of heat generation and the pure air of life, termed 'pneuma'. The atria or auricles were mentioned as soft hollow structure around the ventricles in close proximity to the origins of vessels but were not recognised as part of the heart [2].

The seminal work of Leonardo da Vinci in clearly demarcating the four chambers, arteries and veins in his sketches paved the way for contemporary cardiac anatomy as we know it today. His work distinguished the atria and auricles as separate entities, though it should be mentioned that he referred to the atria as 'upper ventricles' and to the ventricles as 'lower ventricles', with the auricles as 'ears' or 'auricular appendages' [2]. Significantly later, these innovations were expanded upon by William Harvey, whose contributions include identifying the moderator band of the right ventricle and providing a comprehensive description of the previously poorly understood valvular apparatus within the heart [2].

1.2 Atria

1.2.1 Right Atrium

In the frontal plane, the right atrium (RA) is to the right and slightly anterior to the left atrium (LA). The most prominent muscle of the RA is the horseshoe-shaped terminal crest (*crista terminalis*), which marks the line between the endocardial surfaces of the venous component and the wall of the right atrial appendage.

The terminal crest extends from the left side of the superior vena cava (SVC) ostium and goes down laterally to the right of the inferior vena cava (IVC). The terminal crest may act as a physiological barrier to transverse conduction towards the intercaval area, especially in atrial flutter. However, the precise mechanism of this barrier function is yet to be fully established [3].

The sinoatrial node (SAN) is a diffuse collection of cells in the high RA that exhibit automaticity, that is, the capacity to spontaneously depolarise and generate action potentials (APs) [3]. Specifically, the SAN is usually localised within the terminal crest, at its anterolateral junction with the SVC, although in some people there are nodal extensions into the atrial myocardium and/or SVC myocardial sleeves. The intercaval area refers to the right hand of the RA, with the SVC and IVC openings at the top and bottom, respectively. The most proximal part of the SVC is characterised by irregular sleeves of atrial myocardium that extend on the outer side of the venous adventitia. The coronary sinus lies inferior in the RA, close to the right side of the IVC orifice. A crescent-shaped valve, the *thesbian valve*, guards its mouth [3] (◘ Fig. 1.1).

1.2.2 Left Atrium

From the frontal aspect of the chest, the LA is the most posteriorly located of the cardiac chambers. This chamber begins at the pulmonary venoatrial junctions and extends to the fibro-fatty atrioventricular junction at the mitral orifice. The left atrial appendage forms part of the left atrial margin and often overlies the left circumflex coronary artery. Interestingly, the left atrial appendage is the only remnant of the embryonic LA, whereas the smooth-walled LA body develops later.

In the posterior wall of the LA lies the outlets of the four pulmonary veins. Notably, despite the description of the venoatrial junction, the structural border between the vein and atrium is indistinct and atrial tissue extends several centimeters into the sleeves of the pulmonary veins. Haïssaguerre et al. identified these sleeves as an important source of ectopic foci involved in initiating paroxysmal atrial fibrillation (AF) [5]. This will be discussed further in ▶ Chap. 8.

1.2.3 Clinical Implications: Left Atrial Size

LA size increases in the presence of pressure or volume overload and so can be a reliable indicator of left ventricular (LV) diastolic function in people without systolic heart failure or mitral valve disease. LA pressure overload occurs usually with mitral valve disease and LV dysfunction [6]. LA volume overload can be the result of mitral valve regurgitation or high output states such as chronic anaemia. When the left atrial chamber becomes dilated,

■ **Fig. 1.1** Illustrated anatomy of the right atrium, with the heart shown in standard anatomical position. The blue dash shows the crista terminalis. *CS* Coronary Sinus, *IVC* Inferior Vena Cava, *SVC* Superior Vena Cava, *TV* Tricuspid Valve. (Image from [4])

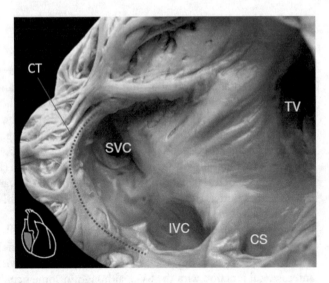

the pulmonary bifurcation may be shifted upward, closer to the aortic arch and potentially compressing the left recurrent laryngeal nerve [6].

1.2.4 Inter-Atrial Septum

The inter-atrial septum comprises the floor of the oval fossa (*fossa ovale*) and the surrounding anterior–inferior myocardium. The most common conduction pathway from the sinus node between the two atria is through *Bachmann's bundle*, also known as the interauricular band. This bundle runs along the anterior left atrial wall, where it merges with superficial and circumferentially arranged myocardial strands of subepicardium. The only place where the bundle appears distinct is at the anterior inter-atrial groove [6].

1.3 Ventricles

The ventricles are characterised by thicker myocardial walls, necessary for the generation of chamber pressures sufficient to eject blood into the pulmonary and systemic circulations by the right and left ventricles, respectively.

1.3.1 Left Ventricle

The LV is thickest at the cardiac base and gradually becomes thinner towards the apex, comprising an endocardial surface with an interwoven network of thin muscle bundles (trabeculations) on a third of the chamber. The geometry of the LV resembles an ellipsoid whose long axis is directed from apex to base. Upon dissection, the LV wall is composed of three layers, namely a superficial epicardium, middle myocardium and deep endocardium, characterised by changes in fibre orientation of the myocardial strands.

As a result of this complex fibre configuration, each ventricular contraction (systole) consists of narrowing, shortening, lengthening, widening, twisting and uncoiling of some or all of the overlapping fibres and different layers in the ventricular wall.

Modelling studies demonstrate that whilst the contraction of exclusively longitudinal or circumferentially arranged myocytes would yield ejection fractions of only 15% and 28%, respectively, the three-dimensional contractile architecture present in the LV facilitates three-dimensional contractions capable of ejection fractions of >60% [7].

1.3.2 Clinical Implications: Cine MRI

Contractile function has traditionally been evaluated using echocardiography and volume-based parameters such as ejection fraction (EF) [8]. Even though echo cardiography affords some assessment of regional wall motion or visual estimation of regional thickening, this is seldom reproducible [8]. Moreover, this method also suffers from a lack of standardisation.

During systole the ventricular myocardium shortens in the longitudinal and circumferential planes and thickens in the radial plane, with reciprocal changes during diastole. Cine MRI is a technique that takes images of the heart throughout the cardiac cycle and displays the cardiac motion in a cine loop. Due to high tissue contrast between the myocardium and blood, changes in the cardiac chamber and ventricular walls can be visualised. Ventricular volume can then be estimated from these images by adding multiple parallel sub-volumes in what is known as the 'short-axis view' [9]. This involves the use of Simpson's rule, which states that 'ventricular volume at a particular cardiac phase can be estimated by adding the encircled areas of a stack of slices from the apex of the heart to the base'.

Once the ventricular volumes are determined at all time frames in a cardiac cycle, a volume–time curve of the ventricle can be made and the peak and trough values used to estimate end-diastolic volume (EDV) and end-systolic volume (ESV), allowing the estimation of ejection fraction. This has proven to be both highly accurate and reproducible [9].

1.3.3 Clinical Implications: Late Gadolinium Enhancement

Late Gadolinium enhancement (LGE) MRI is a technique used to visualise myocardial scarring through the use of water-soluble gadolinium to permeate the interstitial space of tissue. After intravenous injection, both normal and abnormal myocardium display different concentration curves of gadolinium, providing a mechanism of both identifying and evaluating the extent of fibrosis. This is vital for distinguishing ischaemic from non-ischaemic cardiomyopathies, which exhibit distinct enhancement profiles [10]. This is similarly true for different aetiologies of non-ischaemic cardiomyopathy, which also demonstrate discrete enhancement profiles. In patients with ischemic heart disease, distinguishing dysfunctional but viable tissue (referred to as 'hibernating myocardium') from injured myocardium can help to predict the ability to regain contractility following revascularisation therapy.

1.3.4 **Right Ventricle**

The RV is characterised by several interesting anatomical features, appearing as a crescent-shaped chamber comprising a sinus/body and outflow tract/conus, which functions at lower pressure due to possessing one-sixth of the LV muscle mass [11]. The crista supra-ventricular, a U-shaped bundle of longitudinal muscle, divides the two anatomically and functionally distinct areas (Fig. 1.2).

The muscle fibres of the conus run in a parallel alignment from epicardium to endo-cardium, whilst those of the sinus demonstrate a right-angular directional change trans-murally. A parallel fibre arrangement conveys a smaller radius of curvature to the conus, meaning a lower wall stress is generated for any given pressure when compared to the sinus, promoting the absorption and dissipation of sinus-generated pressure [11].

Broadly speaking, the ejection of blood from the RV can be divided into two phases. The first of these is initial ejection, driven by pressure generated from the active shorten-ing of myofibrils and resulting in the propulsion of blood by a reduction in free wall sur-face area and septal-to-free wall distance. Following this is the continued flowing of blood through the outflow tract into the pulmonary trunk, mediated by previously generated momentum [11].

Despite key differences in muscle mass and chamber geometry, both ventricles are bound by spiralling muscle bundles that encircle them to form a functionally single unit. The superficial muscle layers of the sinus are directly continuous with the superficial layers of the LV, whereas fibre bundles in deeper layers are continuous with those of the interven-tricular septum. The intimate relationship of the RV and LV creates a perpetual interplay

 Fig. 1.2 Illustrated anatomy of the right ventricle, with the heart shown in standard anatomical position.
PV pulmonary valve, *RV* right ventricle, *PM* papillary muscle, *MB* moderator band.
(Image from [12])

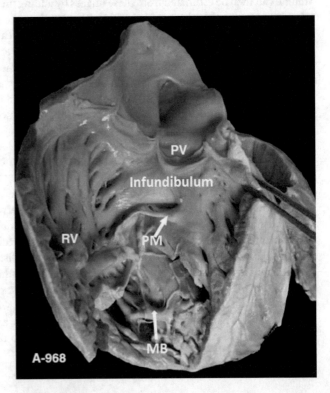

between the two, and as such, the distension of one ventricle can impact upon the other, e.g. LV contraction contributes to the generation of RV systolic pressure [11].

1.4 Great Vessels

The 'great vessels' refers to the five blood vessels that directly interact with the myocardium, namely the:

- Aorta
- Pulmonary arteries (derived from the pulmonary trunk)
- Pulmonary veins
- Superior vena cava (SVC)
- Inferior vena cava (IVC)

The aorta is the largest and most noticeable of the great vessels, arising from the aortic orifice at the base of the LV and progressively becoming the ascending aorta, aortic arch and descending aorta. The ascending portion is the first of these three segments, traversing the pericardium to bring about the coronary arteries that directly perfuse the myocardium. The aortic arch succeeds the ascending branch, distinguished by the branching of three major vessels: the brachiocephalic trunk, left common carotid artery and the left subclavian artery. Finally, the descending aorta permeates the diaphragm into the abdomen [13].

The pulmonary trunk is located anterior and medially to the RA, lying close to the ascending aorta within the pericardium. As it ascends, the pulmonary trunk bifurcates at T5-T6 into the right and left pulmonary arteries that carry deoxygenated blood from the right side of the heart to the lungs. Following gaseous exchange in the alveoli, the pulmonary circulation subsequently returns oxygenated blood to the LA via four pulmonary veins [13].

1.5 Valves

1.5.1 Aortic Valve

Located between the LV and ascending aorta, the aortic valve ensures the unidirectional flow of blood from the LV, optimising coronary blood flow to preserve efficient myocardial functioning. The valve comprises three cusps attached in a semilunar pattern to a crown-shaped annulus. At the basal point of each cusp, the annulus bulges to form three 'pockets' known as the *sinuses of Valsalva*. Functionally, crescent-shaped edges on each cusp alongside the fibrous *nodules of Arantius* form a competent seal that prevents the regurgitation of blood [14].

Before the valve opens, the aortic root prepares to accommodate the large volume of blood exiting from the LV. This involves the expansion of the aortic root and a change in its tilt angle (the angle between the basal and the commissural planes – the end of the closure line). This reduction aligns the left ventricular outflow tract with the ascending aorta and facilitates ejection. During diastole, the tilt angle increases again in order to reduce leaflet stress. During the second half of systole, valve closure is mediated by a conformational change in the three cusps from a conical to a cylindrical shape, with the subsequent deceleration of blood flow accompanied by a drop in the velocity gradient and pressure across the valve [14].

1.5.2 Pulmonary Valve

The second of the semilunar valves, the pulmonary valve similarly comprising three leaflets. The base of each leaflet is supported by the right ventricular infundibulum, with flow through the pulmonary valve exhibiting a similar profile to the aortic valve, albeit with lower velocity and thus pressure [15]. The valve opens during systole and closes at the end of the deceleration phase of systole.

1.5.3 Mitral Valve

Sitting between the LA and LV, the mitral valve consists of four major components: leaflets, chordae tendinae, papillary muscles and the valve annulus. The two mitral valve leaflets insert into the papillary muscles and are continuous with one another, being connected at the commissures. The anterior leaflet, also known as the aortic leaflet, has a rounded free edge and lies in fibrous continuity with the two leaflets of the aortic valve. In contrast, the posterior leaflet hangs like a curtain to line the remainder of the valve circumference [16].

The free edge of the mural leaflet is further divided into three segments: the lateral, middle and medial scallops. The chordae tendinae insert into the under surfaces of both leaflets and have their origin in either two groups of papillary muscles or the posteroinferior LV wall itself [16]. These papillary muscles normally arise from the apical and middle thirds of the LV and as such any alteration in the geometry of the left ventricle can affect valvular function.

During isovolumic relaxation, pressure in the LA exceeds that of the LV, inducing the opening of the mitral valve and subsequent ejection of blood. Following active relaxation, fluid deceleration initiates partial mitral valve closure, which is enhanced by atrial systole [17].

1.5.4 Tricuspid Valve

Similar to the mitral valve, the tricuspid valve comprises three leaflets: anterior, posterior and septal, all of which are attached to a fibrous annulus less easily identifiable than that of the mitral valve [18]. Attached to each leaflet is an eponymous papillary muscle and its paired chordae tendinae. It is noteworthy that blood flow through the tricuspid valve occurs at significantly lower velocities than through the mitral, facilitated by both lower ventricular pressure generation and also a notably larger valve orifice. The tricuspid valve also opens earlier than the mitral due to the shorter RV isovolumic relaxation time [18] (◘ Fig. 1.3).

1.6 Coronary Vasculature

During early cardiac development, the walls of the heart are extensively trabeculated, ensuring that the flow of blood through the primitive heart tube adequately perfuses the underlying endocardium and myocardium. This is later supplanted by a coronary vascu-

▣ Fig. 1.3 Illustrated anatomy of the tricuspid valve inlet and leaflets. The solid line marks the attachment of the tricuspid valve. The dotted line marks to tendon of Todaro, *AS* Anteroseptal leaflet, *S* septal leaflet, *I* inferior papillary muscle, *CS* coronary sinus, *OF* oval fossa. (Image from [19])

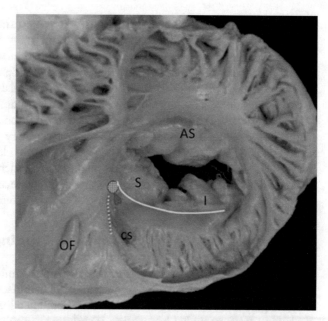

lature consisting of a left coronary artery (LCA) that subsequently bifurcates into left anterior descending (LAD) and circumflex branches, in addition to a right coronary artery (RCA) that gives rise to right marginal, sinoatrial and infundibular branches [20].

Left Coronary Artery It originates from the left sinus of Valsalva and travels anteriorly and to the left, ultimately dividing into the circumflex and anterior descending arteries. These branches supply most of the LV, LA and ventricular septum [20].

Left Anterior Descending Artery A direct continuation of the LCA, it emerges behind the pulmonary trunk to travel down the anterior interventricular groove, towards the apex. The LAD then ascends within the inferior interventricular sulcus to give off branches that supply the apical surfaces of both ventricles' inferior walls. Importantly, the LAD is the origin of two groups of vessels: diagonal arteries that supply the parietal wall of the left ventricle, and septal perforators that perfuse the anterior two thirds of the muscular ventricular septum [20].

Circumflex Artery It branches at a perpendicular angle to the main stem to traverse the left atrioventricular groove. The vessel then gives off branches that supply the posterior and lateral walls of the LV, in addition to the superolateral papillary muscle. Smaller unnamed branches arise from the circumflex artery as it travels along the atrioventricular groove to supply the root of the aorta and ventricular myocardium adjacent to the groove [20]. In one tenth of individuals, the circumflex gives rise to the artery that supplies the atrioventricular node [20].

Right Coronary Artery It arises from the right sinus of Valsalva to descend anteriorly and inferiorly down the atrioventricular groove, giving rise to several branches, the first of which is the infundibular artery that supplies the right infundibular musculature. Second is the sinoatrial artery responsible for perfusion of the sinus node, followed by the right marginal artery [20].

1

1.6.1 Clinical Implications: Coronary Stenosis

Despite significant therapeutic advancements, coronary artery disease (CAD) remains one of the world's foremost mortality burdens, presenting as stenosis of the coronary arteries as a result of arteriosclerosis, insufficient myocardial perfusion (angina pectoris) or total occlusion, and leading to necrosis of the myocardium (myocardial infarction/MI) [21]. The correlation between the degree of anatomical coronary stenosis and reduction in blood flow provides a foundation for the use of coronary CT angiogram to diagnose obstructive CAD. CT utilises multislice technology to provide high quality 3D imaging of the vessels [22]. Unlike invasive coronary angiography, which allows for quantification of dynamic flow in coronary arteries, conventional CT coronary angiography is unable to characterise the haemodynamic significance of coronary stenosis.

1.7 Where We're Heading: Cardiac 3D Printing

'Three-dimensional (3D) printing' is a manufacturing technique that uses the deposition of successive layers of a structure, termed 'additive manufacturing', to create 3D objects from a digital file [23]. In response to the aforementioned limitations of existing imaging modalities, previous studies have shown 3D printing as a suitable mechanism for visualising geometrically complex regions of tissue and/or vasculature through the creation of rapid, high-fidelity anatomical models [24, 25].

Indeed, the capacity of this technique to manufacture heterogeneous and elaborate structures identifies it as a potentially valuable tool for the planning of congenital heart disease surgeries, as demonstrated in the cases shown by Farooqi et al., who concluded that 'considering that the overall response to this technology from the medical community, especially surgeons, has been positive, the main barriers to more widespread use are largely technical' [26].

Take-Home Message

- The heart is composed of four chambers: two atria and two ventricles, which are segregated by septa to create two independent (left- and right-sided) circulations.
- The ejection of blood relies on the timed opening and closing of two semilunar and two atrioventricular valves.
- Blood is ejected into either the aorta or pulmonary trunk – two of the four great vessels of the heart. From the aorta, a proportion of the blood returns to supply the heart via the left and right coronary arteries.
- Three-dimensional (3D) printing is a novel manufacturing technique that may hold clinical value through the creation of high-fidelity models of complex anatomical structures.

References

1. Jensen B, Wang T, Christoffels VM, Moorman AFM (2013) Evolution and development of the building plan of the vertebrate heart. Biochim Biophys Acta, Mol Cell Res 1833:783–794

2. Loukas M, Youssef P, Gielecki J, Walocha J, Natsis K, Tubbs RS (2016) History of cardiac anatomy: a comprehensive review from the Egyptians to today. Clin Anat 29(3):270–284
3. Corradi D, Maestri R, MacChi E, Callegari S (2011) The atria: from morphology to function. J Cardiovasc Electrophysiol 22(2):223–235
4. Zhu X (2015) The Right Atrium. In: Zhu X. (eds) Surgical Atlas of Cardiac Anatomy. Springer, Dordrecht
5. Haïssaguerre M, Jaïs P, Shah DC, Takahashi A, Hocini M, Quiniou G et al (1998) Spontaneous initiation of atrial fibrillation by ectopic beats originating in the pulmonary veins. N Engl J Med 339(10): 659–666
6. Ho SY, Cabrera JA, Sanchez-Quintana D (2012) Left atrial anatomy revisited. Circ Arrhythmia Electrophysiol 5(1):220–228
7. Kocica MJ, Corno AF, Carreras-Costa F, Ballester-Rodes M, Moghbel MC, Cueva CNC et al (2006) The helical ventricular myocardial band: global, three-dimensional, functional architecture of the ventricular myocardium. Eur J Cardiothorac Surg 29:S21
8. Kotecha D, Mohamed M, Shantsila E, Popescu BA, Steeds RP (2017) Is echocardiography valid and reproducible in patients with atrial fibrillation? A systematic review. Europace 19:1427
9. Stucky DJ, Carr CA, Tyler DJ, Clarke K (2008) Cine-MRI versus two-dimensional echocardiography to measure in vivo left ventricular function in rat heart. NMR Biomed 21:765
10. Doltra A, Amundsen B, Gebker R, Fleck E, Kelle S (2013) Emerging concepts for myocardial late gadolinium enhancement MRI. Curr Cardiol Rev 9:185
11. Spicer DE, Anderson RH (2013) Methodological review of ventricular anatomy – the basis for understanding congenital cardiac malformations. J Cardiovasc Transl Res 6:145–154
12. Zhu X (2015) The Right Ventricle. In: Zhu X. (eds) Surgical Atlas of Cardiac Anatomy. Springer, Dordrecht
13. URBAN Fisch (2011) General anatomy and musculoskeletal system. In: Sobotta, atlas of human anatomy
14. Chester AH, El-Hamamsy I, Butcher JT, Latif N, Bertazzo S, Yacoub MH (2014) The living aortic valve: from molecules to function. Glob Cardiol Sci Pract 2014(1):11
15. Bateman MG, Hill AJ, Quill JL, Iaizzo PA (2013) The clinical anatomy and pathology of the human arterial valves: implications for repair or replacement. J Cardiovasc Transl Res 6(2):166–175
16. McCarthy KP, Ring L, Rana BS (2010) Anatomy of the mitral valve: understanding the mitral valve complex in mitral regurgitation. Eur. J. Echocardiogr 11:i3–9
17. Sacks MS, Yoganathan AP (2007) Heart valve function: a biomechanical perspective. Philosop Transact Royal Soc B: Biol Sci 362:1369–1391
18. Barker TA, Wilson IC (2011) Surgical anatomy of the mitral and tricuspid valve. In: Mitral valve surgery, pp 3–19
19. McCarthy KP, Robertus JL, Yen Ho S (2018) Anatomy and Pathology of the Tricuspid Valve. In: Soliman O.I., ten Cate F.J. (eds) Practical Manual of Tricuspid Valve Diseases. Springer, Cham
20. Loukas M, Sharma A, Blaak C, Sorenson E, Mian A (2013) The clinical anatomy of the coronary arteries. J Cardiovasc Transl Res 6(2):197–207
21. Davies SW (2001) Clinical presentation and diagnosis of coronary artery disease: stable angina. Br Med Bull 59:17
22. Nasis A, Meredith IT, Cameron JD, Seneviratne SK (2015) Coronary computed tomography angiography for the assessment of chest pain: current status and future directions. Int J Cardiovasc Imaging 31(Suppl 2):125–143
23. Vukicevic M, Mosadegh B, Min JK, Little SH (2017) Cardiac 3D printing and its future directions. JACC Cardiovasc Imaging 10:171–184
24. Markert M, Weber S Lueth T C and IEEE 2007 a beating heart model 3D printed from specific patient data annual int. conf. In: IEEE engineering in medicine and biology society 1-162007, pp 4472–4475
25. Shiraishi I, Kurosaki K, Kanzaki S, Ichikawa H (2014) Development of super flexible replica of congenital heart disease with stereolithography 3D printing for simulation surgery and medical education. J Cardiac Failure 20:S180–S181
26. Farooqi KM, Saeed O, Zaidi A et al (2016) 3D printing to guide ventricular assist device placement in adults with congenital heart disease and heart failure. JACC Heart Fail 4:301–311

Early Mechanisms of Cardiac Development

Jack Griffiths and Thomas Brand

© Springer Nature Switzerland AG 2019
C. Terracciano, S. Guymer (eds.), *Heart of the Matter*, Learning Materials in Biosciences,
https://doi.org/10.1007/978-3-030-24219-0_2

2

What You Will Learn in This Chapter

This chapter initially introduces the importance of animal models in understanding cardiac development. We will discuss the early stages of embryonic development, including gastrulation, formation of the linear heart tube and the looping process. We will then explore the concept of the organiser during early vertebrate development using the example of the Spemann-Mangold organiser of the amphibian embryo. Finally, we will expand on how cells in the embryo are able to interpret morphogen gradients by introducing the French Flag Model.

Learning Objectives

- Discuss the early stages of cardiac development, including the migration of cells to the lateral plate mesoderm and formation of the linear tubular heart.
- Understand the organiser principle using the example of the Spemann-Mangold organiser in the amphibian embryo and describe some key mediators involved in this process.
- Appreciate the relevance of the French Flag Model in understanding the importance of morphogen gradients.

2.1 An Introduction to Cardiac Developmental Biology

Cardiac developmental biology represents an ever-evolving and conceptually challenging field, with many areas still poorly understood. A nexus of simultaneous processes is required for the formation of this morphologically and functionally complex organ, beginning with the migration of precardiac mesoderm cells in the gastrula to the lateral plate mesoderm, the induction of cardiac mesoderm and the formation of bilateral endocardial tubes that eventually coalesce to form the linear heart tube [1].

From here, the embryonic heart undergoes vast electrical and architectural changes, whilst simultaneously experiencing an evolutionarily conserved looping process that results in a rearrangement of the relative positions of the forming atrial and ventricular chambers, the convergence of the venous and arterial poles and ultimately generating the topography of the mature heart [1].

2.2 Comparing Animal Models of Cardiac Development

Much of our current understanding of cardiac development, both anatomical and molecular, derives from studies utilising a variety of animal models [2]. While no organism optimally replicates the development of the human heart, each of the four major animal models possesses characteristic advantages and limitations, highlighted in the table below [1–3]:

	Advantages of model	Limitations of model
Zebrafish	1. Fully sequenced genome 2. Short generation time 3. Transparent embryos 4. Transgenesis is possible (can incorporate genes from external sources)	1. Manipulations (transplantation of cells) is difficult 2. Tetraploid genome

	Advantages of model	Limitations of model
African claw frog	1. Excellent for experimental manipulations 2. Transgenesis is possible 3. Fully sequenced genome	1. Genetic manipulation is difficult (too many gene knockouts required) 2. Long generation time
Chick	1. Experimental manipulations possible 2. Non-aquatic embryo 3. Established embryological model 4. Fully sequenced genome	1. Genetic manipulation is difficult and expensive
Mouse	1. Developmental pathway similar to humans 2. Fully sequenced genome 3. Transgenesis is possible 4. Highly suitable to model human congenital heart disease	1. Early stages of heart development are difficult to study 2. Much higher heart rate and therefore differences in cardiac physiology

2.3 Gastrulation

Without delving excessively into the early stages of embryonic development, it is first important to discuss *gastrulation*: a vital process that occurs in most mammals following the formation of the blastula (usually around day 16 in the human embryo) [1] and preceding the development of organs (*organogenesis*). Necessary for the commencement of gastrulation is the appearance of the *primitive streak*, which demarcates the embryonic disc into anterior and posterior divisions through the creation of the anteroposterior axis. At this stage, the embryonic disc comprises the epiblast (which gives rise to the embryo) and the hypoblast (participates in the formation of extraembryonic tissues) [1]. During gastrulation, epiblast cells ingress through the primitive streak whilst simultaneously undergoing epithelial-to-mesenchymal transition, resulting in the formation of three germ layers:

- *Ectoderm*: outermost layer
- *Mesoderm*: middle layer
- *Endoderm*: innermost layer

This migratory process is orchestrated by a complex network of signalling molecules released by the primitive streak, including Nodal, BMP4 and Wnt [4]. Following cell migration and the completion of gastrulation, three key axes have been established in the embryo: anteroposterior, dorsoventral and left-right [5]. These embryonic axes are vital for the correct positioning of the organ primordia, achieved through the generation of concentration gradients of a cocktail of signalling molecules at any given site. Following gastrulation, four key subsets of mesoderm emerge along the dorsoventral axis [5, 6]:

- *Chorda Mesoderm*: the most medial variant and progenitor of the notochord – a cartilaginous, flexible rod required for muscle attachment.
- *Paraxial Mesoderm*: adjacent to the chorda mesoderm and gives rise to the head mesoderm and the somites, which are the building blocks of the vertebral column and skeletal muscle.

2

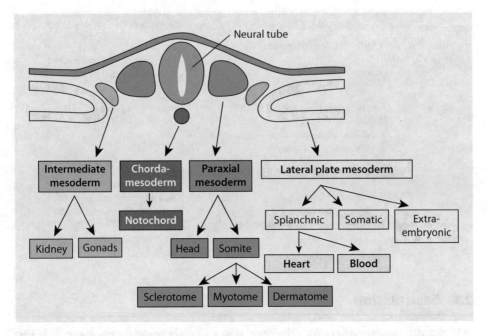

◻ Fig. 2.1   Illustration of the four key mesodermal subsets that arise along the dorsoventral axis following gastrulation. (Adapted from [6])

— *Intermediate Mesoderm*: further lateral, responsible for kidney and gonad formation.
— *Lateral Plate Mesoderm (LPM)*: the most lateral mesoderm that splits into two layers: the splanchnic and the somatic LPM, with the former giving rise to the heart and blood (◻ Fig. 2.1).

2.3.1 French Flag Model of Morphogen Expression

A morphogen is a paracrine-acting signalling molecule whose precisely controlled expression governs cellular responses specific to the developmental process [1]. As previously alluded to, both agonist and antagonistic signalling molecules exist, resulting in the formation of concentration gradients.

The degree of ligand-bound receptors ultimately determines the strength of signal transduction, in itself leading to the activation or repression of target genes via transcription factor interactions [1]. The morphogen diffuses from a localised source to manifest as a concentration gradient across a developing tissue [7]. The *French Flag Model* (◻ Fig. 2.2) helps to illustrate this [1]. High morphogen concentrations induce the activation of a different set of genes to those activated at low concentration [7].

2.4 Induction and the Organiser Principle

The importance of local signalling in embryonic development was elegantly demonstrated in 1924 by Hans Spemann and Hilde Mangold, who transplanted a dorsal blastopore lip extracted from one newt species under the ectoderm of a different newt species [8].

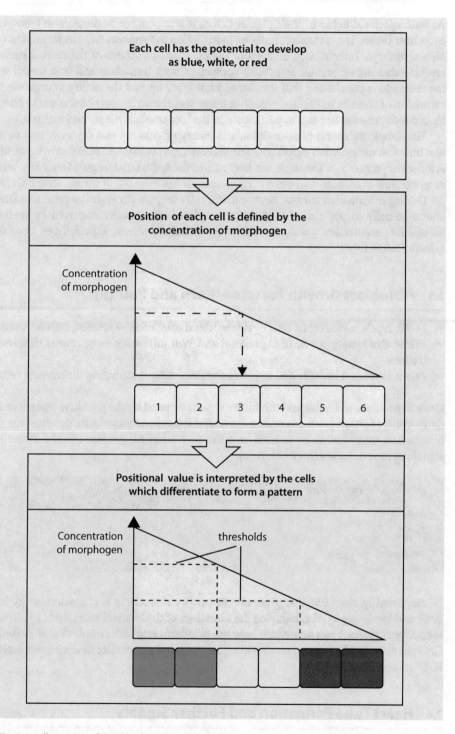

Fig. 2.2 Illustration of the *French Flag Model*, whereby a given morphogen manifests along a concentration gradient, with each concentration inducing a different set of genes. This results in the formation of three different cell types, represented by the colours blue, white and red. (Image from [1])

2

As both species differed in their pigmentation, it was possible to distinguish transplant from host tissue. The transplant induced gastrulation movements and the formation of a second embryo. Interestingly, analysis of the pigmentation status of the second embryo revealed that the embryonic structures comprised both transplant and host tissue, with the researchers concluding that the dorsal blastopore lip had the ability to organise the formation of an embryo [8]. The organiser tissue was able to instruct host tissue to change their developmental fate and to participate in the formation of the second embryo.

The ectopic induction of gastrulation and embryo formation also demonstrated for the first time that mesodermal signals influenced central nervous system (CNS) development in ectodermal progenitors. The organiser, later named the Spemann-Mangold Organiser, serves as an example of *induction*, whereby certain cells influence the fate of surrounding cells [8].

During mammalian cardiac development, cells migrate through the primitive streak antero-laterally to populate the lateral plate splanchnic mesoderm, driven by spatially localised concentration gradients within the primitive streak, especially of fibroblast growth factors [5].

2.5 Fibroblast Growth Factors – 'Push and Pull' [5]

- *FGF8*: repels mesodermal cells → drives lateral migration → sending cells outward.
- *FGF8*: also synergises BMP expression and Wnt inhibition → promotes differentiation.
- *FGF4*: expressed laterally and attracts migratory cells → attracting the moving cells.

Once in position, inductive and inhibitory signals secreted by the primitive foregut endoderm initiate further gradients within the lateral splanchnic mesoderm, determining differentiation into cardiac mesoderm and the formation of the heart fields. Some key signalling molecules are listed below [5]:

Endoderm-derived inducers	Mesodermal inhibitory signals
BMP	BMP antagonists (chordin and noggin)
Wnt antagonists, e.g. crescent	Wnt (canonical)
FGF8	
Activin	

Importantly, the LPM that gives rise to cardiac mesoderm is characterised by high BMP and low levels of Wnt, inducing the formation of the bilateral heart fields [5]. By this induction logic, one can appreciate how the ectoderm and notochord, vital sites for the CNS but not heart formation, are rich in signals inhibitory to cardiac development, namely BMP antagonists and Wnt.

2.6 Heart Tube Formation and Further Signals

Cardiac mesoderm induction of the splanchnic LPM results in the formation of heart fields on either side of the neural tube. The eventual ramification of further foregut endoderm-derived signalling is the formation of a myocardial layer and endocardial tubes. In *Drosophila*, heart formation is regulated by the *tinman* gene, which encodes a transcription factor specific to cardiac progenitor cells whose mutation is associated with the total absence of a heart tube [9].

Nkx2.5 is the vertebrate homologue of tinman and is also expressed in the heart from the time of cardiac mesoderm formation [10]. Heart development in the *Nkx2.5* knockout mutant is abnormal, demonstrating an essential role of Nkx2.5 in vertebrate heart development [10]. That said, *Nkx2.5* is of lesser importance than *tinman* is in Drosophila, evidenced by the heart developing until the looping stage has been reached [11]. The milder phenotype of the *Nkx2.5* knockout mutant is probably due to the interplay of multiple factors in the effectuation of vertebrate cardiac mesodermal specification, which are able to compensate the loss of *Nkx2.5*. Nonetheless, Nkx2.5 is a key marker of cardiac mesoderm in vertebrates [10, 11].

Concurrent with, and partly facilitated by, gut tube fusion, the two endocardial tubes are pushed into the thoracic cavity, where they undergo apoptosis-mediated fusion to form a single heart tube [1]. The newly formed heart tube remains attached to the dorsal pericardial cavity by dorsal mesocardium, the remnants of which are the transverse and oblique pericardial sinuses visible in the mature myocardium. Tubular heart formation can be thought of as having passive and active components, with gut tube formation passively initiating heart field fusion via a process actively coordinated by myriad genes.

2.7 Second Heart Field

By now, the primitive heart is a single tube comprising cardiac mesoderm of the first heart field (FHF). Elongation of the heart tube is driven by a second cardiac mesoderm population, which migrates into the heart to add myocardial cells at the cranial and caudal poles of the primary heart tube. This second heart field (SHF) population is located medial to the FHF mesoderm in the gastrula stage embryo [12].

The importance of the SHF is highlighted by its contributions to the atria, right ventricle (RV), and outflow tract (OFT). Evidence of the latter was provided first by de la Cruz et al., who observed that OFT formation occurred after the formation of the primary heart tube [13]. Further research identified islet-1 (ISL-1) as a transcription factor that is specifically expressed in the cells of the SHF [14]. ISL-1-expressing SHF cells added to the cranial pole of the embryonic heart differentiate into outflow tract myocardium, with further investigations by Rothenberg et al. revealing that much of this SHF-derived outflow tract myocardium is later lost through apoptosis and replaced by aortic smooth muscle originating from neural crest cells [15].

Unlike FHF cells, which require high BMP levels in order to differentiate, express contractile proteins and undergo primary heart tube formation, SHF cells are prevented from prematurely differentiating, probably achieved through exposure to BMP antagonists and canonical Wnt signals. The SHF cells are responsible for the elongation of the primary heart tube at either pole [12]. SHF-derived cardiac tissue can be identified using ISL-1 as a lineage marker, which is only very transiently expressed by FHF cells [12, 16]. ◘ Figure 2.3 broadly illustrates some of the cardiac regions derived from SHF cells [16].

2

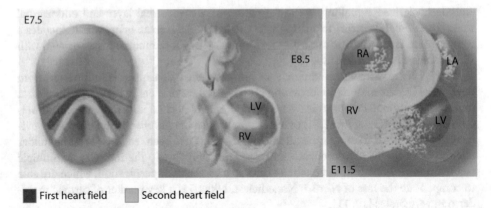

☐ **Fig. 2.3** Illustration of second heart field (SHF) cells at three stages of cardiac development. Myocardial regions derived from first and second heart fields are shown in red and green, respectively. Note several key areas that originate from SHF cells, including the primitive OFT, the right ventricle and part of the atria. (Image modified from [15])

2.7.1 Inductive Signalling for the FHF and SHF

HAND2 is a cardiac transcription factor expressed in the RV. Interestingly, loss of the *Hand2* gene in the mouse results in the absence of RV and OFT tissue. This suggests an association between HAND2 and the formation or differentiation of SHF progenitors.

Transcription Factors Controlling FHF Development   GATA4 drives Tbx5 and Nkx2.5 production, promoting cardiac-specific differentiation and the formation of the tubular heart [17]. Tbx5 deficiency therefore affects atrial and LV myocardium while OFT and RV tissue is normal.

Genes Essential for the SHF Lineage   Hand2 and FGF10 are essential for the SHF lineage.

2.8 Cardiac Looping

The now-elongated, nearly symmetrical heart tube remains attached to the dorsal pericardial cavity via dorsal mesocardium at the venous and arterial poles, however the non-polar dorsal myocardium is lost by apoptosis [18]. Elongation of the heart tube via SHF cell recruitment therefore leads to a ventral bulging of the growing heart. Simultaneous with this is a twisting of the heart tube via rotation at either pole. In an evolutionarily conserved process, looping begins with the growing heart tube bending towards the right, marking the beginning of asymmetric development of the heart [18, 19]. During looping, the atria are pushed in a cranial direction and are located in a dorsoanterior position relative to the ventricles by the end of looping [18]. This process is also important for the alignment of the inflow and outflow tracts and is closely orchestrated by a concert of signalling molecules, ultimately leading to the formation of the four-chambered heart (☐ Fig. 2.4).

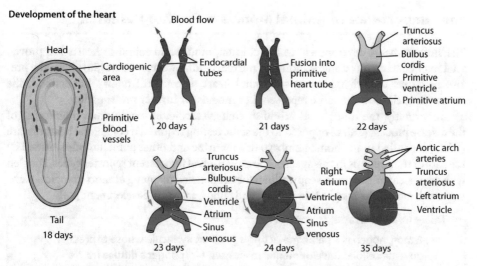

Development of the heart

○ **Fig. 2.4** Illustration of the broad anatomical changes that occur during heart development, including endocardial tube fusion and cardiac looping. (Image from [19])

2.9 What We Don't Know: Valveless Unidirectional Blood Flow

Prior to looping, the embryonic tubular heart is characterised by slow, inefficient, unidirectional contractions that propel blood from the inflow to outflow tracts, into the developing vasculature. Two theories have been proposed as to the mechanism that achieves unidirectional blood flow in the valveless early primitive heart: [1] the heart is a peristaltic pump, and [2] the early embryonic heart is a Liebau pump [20].

The first theory suggests that the tubular heart contracts in a peristaltic manner, with a series of wave-like muscular contractions originating from the sinus venosus (near the inflow tract) propelling blood through the cylindrical chamber. Indeed, this is supported by observational studies and is paralleled in other organ systems, e.g. the gastrointestinal tract [20, 21].

The alternative theory proposes that the early embryonic heart achieves directional blood flow via a phenomenon first by Gerhart Liebau in 1954 [22]. The 'Liebau effect' suggests that the periodic compression of a single site within the heart tube creates an impedance pump to propel blood through the outflow tract [22, 23]. Scientifically, this phenomenon can be observed in any compliant tube and is commonly exploited in microfluidic engineering.

Despite several simulation experiments by Hiermaier and Männer in an attempt to clarify which mechanism occurs in the embryonic heart, many questions persist [20, 24]. That said, the investigations did yield evidence that the looped heart exhibits improved pumping efficiency compared to a linear configuration [20].

As previously discussed, the study of cardiac development is reliant upon the utilisation of animal models. Indeed, based on descriptive zebrafish studies, Forouhar et al. concluded that Liebau's impedance theorem is the more likely model for how the embryonic heart pumps blood [25]. Such is the nature of the controversy however, that this conclusion was discredited by proponents of the peristaltic model based on the tenet that the results could only be compared to a technical peristaltic pump, and not a biological peristaltic [20, 26, 27].

2

2.10   Importance of Animal Models in Further Research

The study of heart development in the chicken, mouse and zebrafish hearts has proven vital in enhancing our understanding of normal cardiac development [28]. With the precise pathways underlying many congenital heart defects still poorly understood, the mouse and zebrafish therefore represent key models for further investigation.

Importantly, the transparent zebrafish embryo facilitates longitudinal monitoring of the disease process [28], as opposed to the static testing required if congenital heart disease (CHD) is studied in the human embryo or in murine and other animal models. Aided by the ease of transgenesis in this species, the utilisation of fluorescent reporter genes, coupled with advancements in microscopy, allows for the detailed mapping of morphological pathways, enhancing our insight into normal and abnormal heart development [28].

> **Take-Home Message**
>
> — A morphogen is a paracrine-acting signalling molecule whose expression governs cellular developmental responses. Morphogens diffuse from a localised source to manifest along a concentration gradient in a developing tissue.
> — Cells migrate through the primitive streak antero-laterally. Lateral plate mesoderm gives rise to cardiac mesoderm and is characterised by high BMP and low Wnt signalling.
> — Second heart field (SHF) cells migrate into the primary heart tube at the cranial and caudal poles, contributing to the primitive OFT, right ventricle and part of the atria.
> — Two theories exist for the mechanism of unidirectional blood flow in the valveless tubular heart: [1] the early heart is a peristaltic pump, and [2] the early heart is a Liebau pump.

References

1. Wolpert L, Tickle C (2011) Principles of development, vol 616, 4th edn. Oxford University Press, Oxford
2. Bakkers J (2011) Zebrafish as a model to study cardiac development and human cardiac disease. Cardiovasc Res 91:279
3. Wittig J, Münsterberg A (2016) The early stages of heart development: insights from chicken embryos. J Cardiovasc Dev Dis 3:12
4. Zaffran S, Frasch M (2002) Early signals in cardiac development. Circ Res 91:457
5. Brand T (2003) Heart development: molecular insights into cardiac specification and early morphogenesis. Dev Biol 258:1
6. Gilbert S, F, Bareresi MJF (2016) Developmental biology, 11th edn. Sinauer, Sunderland, MA, USA. 888 p. 444
7. Wolpert L (1969) Positional information and the spatial pattern of cellular differentiation. J Theor Biol 25:1
8. Niehrs C (2004) Regionally specific induction by the Spemann-Mangold organizer. Nat Rev Genet 5:425
9. Bodmer R (1993) The gene tinman is required for specification of the heart and visceral muscles in Drosophila. Development 118:179
10. Benson DW, Silberbach GM, Kavanaugh-McHugh A, Cottrill C, Zhang Y, Riggs S et al (1999) Mutations in the cardiac transcription factor NKX2.5 affect diverse cardiac developmental pathways. J Clin Invest 104:1567

11. Lyons I, Parsons LM, Hartley L, Li R, Andrews JE, Robb L et al (1995) Myogenic and morphogenetic defects in the heart tubes of murine embryos lacking the homeo box gene Nkx2-5. Genes Dev 9:1654

12. Dyer LA, Kirby ML (2009) The role of secondary heart field in cardiac development. Dev Biol 336:137

13. de la Cruz MV, Gmez CS, Arteaga MM, Arguello C (1977) Experimental study of the development of the truncus and the conus in the chick embryo. J Anat 123:661

14. Cai CL, Liang X, Shi Y, Chu PH, Pfaff SL, Chen J et al (2003) Isl1 identifies a cardiac progenitor population that proliferates prior to differentiation and contributes a majority of cells to the heart. Dev Cell 5:877

15. Rothenberg F, Hitomi M, Fisher SA, Watanabe M (2002) Initiation of apoptosis in the developing avian outflow tract myocardium. Dev Dyn 223:469

16. Parmacek MS, Epstein JA (2005) Pursuing cardiac progenitors: regeneration redux. Cell 120:295

17. Bruneau BG (2013) Signaling and transcriptional networks in heart development and regeneration. Cold Spring Harb Perspect Biol 5:a008292

18. Männer J (2009) The anatomy of cardiac looping: a step towards the understanding of the morphogenesis of several forms of congenital cardiac malformations. Clin Anat 22:21

19. Young KA, Wise JA, DeSaix P, Kruse DH, Poe B, Johnson E, et al. Anatomy & physiology. 1st ed. OpenStax College; 2013. 1335 p

20. Hiermeier F, Männer J (2017) Kinking and torsion can significantly improve the efficiency of valveless pumping in periodically compressed tubular conduits. Implications for understanding of the form-function relationship of embryonic heart tubes. J Cardiovasc Dev Dis 4:19

21. Jaffrin MY, Shapiro AH (1971) Peristaltic pumping. Annu Rev Fluid Mech 3:13

22. Liebau G (1954) Über ein ventilloses Pumpprinzip. Naturwissenschaften 41:327

23. Liebau G (1955) Herzpulsation und Blutbewegung. Z Gesamte Exp Med 125:482

24. Männer J, Wessel A, Yelbuz TM (2010) How does the tubular embryonic heart work? Looking for the physical mechanism generating unidirectional blood flow in the valveless embryonic heart tube. Dev Dyn 239:1035

25. Forouhar AS, Liebling M, Hickerson A, Nasiraei-Moghaddam A, Tsai HJ, Hove JR et al (2006) The embryonic vertebrate heart tube is a dynamic suction pump. Science 312:751

26. Taber LA, Zhang J, Peruccchio R (2007) Computational model for the transition from peristaltic to pulsatile flow in the embryonic heart tube. J Biomech Eng 129:441

27. Baird A, King T, Miller LA (2012) Numerical study of scaling effects in peristalsis and dynamic suction pumping. In: Proceedings of the AMS, special session on biological fluid dynamics: modeling, computations, and applications, New Orleans, pp 129–148

28. Tu S, Chi NC (2012) Zebrafish models in cardiac development and congenital heart birth defects. Differentiation 84:4

Later Mechanisms of Cardiac Development

Beth Taylor and Thomas Brand

© Springer Nature Switzerland AG 2019
C. Terracciano, S. Guymer (eds.), *Heart of the Matter*, Learning Materials in Biosciences,
https://doi.org/10.1007/978-3-030-24219-0_3

What You Will Learn in This Chapter

This chapter expands upon selected processes occurring following the establishment of the first and second heart fields, previously discussed in ▶ Chap. 2. We will begin by exploring the establishment of the left-right axis, both in terms of ionic currents involved, the relevant signalling mediators and any mechanical interplay. We will then examine the development of the cardiac pacemaker, beginning from the electrical activity in the tubular heart, and its progressive compartmentalisation to the sinoatrial (and potentially atrioventricular) nodes. Finally, after analysing the valvular architecture of the heart and origin of the epicardium, we will conclude by briefly discussing the potential utilisation of ectopic Tbx18 induction to generate a biological pacemaker.

Learning Objectives
- Understand the mechanisms of left-right axis formation, including the relevant signalling molecules.
- Be able to discuss the evolution of cardiac pacemaking and its compartmentalisation.
- Consider the employment of ectopic Tbx18 expression in non-pacemaking cardiomyocytes to produce a biological pacemaker.

3.1 Establishing the Left-Right Axis for Asymmetry

Studies in the chick embryo have proven vital in the understanding of how the embryonic heart establishes its left-right asymmetry, vital for cardiac looping and other asymmetries in the mature heart. This process relies on Hensen's node: the avian equivalent of the Spemann-Mangold Organiser found in the amphibian embryo as previously discussed in ▶ Chap. 2. As in other species, Hensen's node serves as a signalling centre orchestrating gastrulation, and is also involved in the establishment of the three main embryonic axes (anteroposterior, dorsoventral and left-right). Two main mechanisms are ultimately responsible:
1. Asymmetric gene expression: the creation of spatially localised morphogen concentration gradients. Remember the 'French Flag Model' used in the previous chapter.
2. Cilia on cells forming the ventral node creating directional morphogen flow.

In the chicken embryo, differential gene expression on either side of Hensen's node creates molecular asymmetry [1]. On the left side of Hensen's node, Sonic hedgehog (Shh) induces the production of Nodal: a member of the transforming growth factor beta (TGF-β) family, which is at this stage of development exclusively expressed on the left side of the embryo. Shh expression is confined to the left side of Hensen's node due to antagonism by the right-sided expression of BMP4 [2]. CFC, a member of the EGF-CFC family of Nodal co-receptors is asymmetrically expressed on the left side and serves to enhance the responsiveness on the left side for Nodal. Partly facilitated by CFC, Nodal signalling induces expression of Lefty in the midline of the node. Lefty acts as a functional antagonist of Nodal signalling, thus isolating Nodal signalling to the left side of Hensen's node (◻ Fig. 3.1).

Aided by Caronte (a Wnt antagonist), Nodal rapidly diffuses laterally from the node through the left paraxial and intermediate mesoderm, into the lateral plate mesoderm. Here, the presence of Nodal induces the expression of Nkx3.2 and Pitx2, both of which negatively feedback on Nodal expression. Enhancing this response in the lateral plate

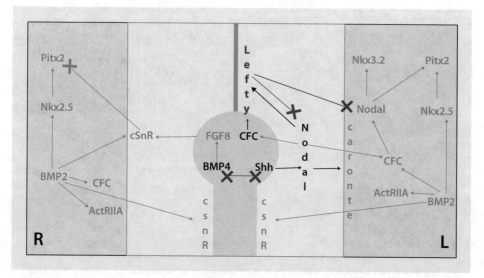

mesoderm are CFC and ActRIIA, both of which are expressed due to the presence of BMP2. With Pitx2 decreasing the concentration of Nodal in the left lateral plate mesoderm, Pitx2 expression is maintained by Nkx2.5, which itself is induced by BMP2. It is Pitx2 that is largely responsible for initiating left-sided morphogenesis, thereby identifying it as an important left-side marker [3] (◻ Fig. 3.2).

On the right side of Hensen's node in avian embryos, BMP4 upregulates the expression of FGF8, responsible for 'driving' cells laterally towards the lateral plate mesoderm. Moreover, FGF8 upregulates cSnR (a member of the Snail family of transcription factors controlling epithelial-mesenchymal transformation (EMT) in the right paraxial mesoderm, with cSnR subsequently inhibiting the Pitx2 homeobox transcription factors, and thereby, the left-sided programme. Despite these differences between the left and right sides, symmetric expression of Nkx2.5, CFC and ActRIIA is present on either side. Yet the question persists about what is establishing the asymmetric expression of BMP4 and Shh on the right and left sides of Hensen's node, respectively. The current consensus supports the view that asymmetric cell migration around the node is involved [4] (◻ Fig. 3.3).

3.1.1 Cilia on the Ventral Nodal Aspect

The other fundamental mechanism to be discussed involves cilia located on the ventral aspect of the node of the mouse, *Xenopus* and zebrafish embryos. Researchers have identified that these cilia are able to rotate clockwise and generate a leftward nodal flow [5]. Surprisingly, this mechanism, which is believed to be of central importance in many vertebrate embryos and possibly also in humans, is not utilised in the chicken embryo, as node cells in this species have short and immobile cilia. Instead, directional migration

3

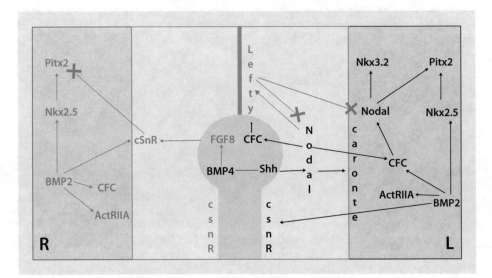

Fig. 3.2 Schematic of the signalling pathways involved in the establishment of the LR axis. With help of Caronte, Nodal diffuses rapidly through the mesoderm while BMP2 establishes the competence of the lateral plate mesoderm to respond to Nodal by inducing CFC and ACTRIIA. Nodal induces Pitx2 and NKX3.2. Nkx.5 is required to maintain the expression of Pitx2 in the LPM. (Modified from [1])

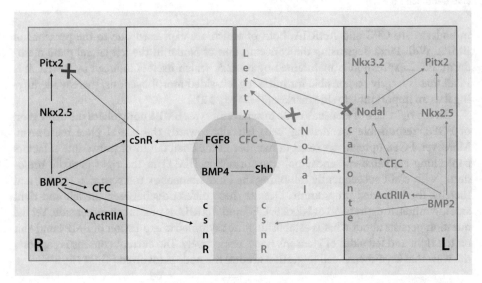

Fig. 3.3 Schematic of the signalling pathways involved in the establishment of the LR axis. This figure depicts key mediators involved in this process, particularly those relevant to right side identity. (Modified from [1])

around Hensen's node, also observed in the porcine embryo, appears to be responsible for the establishment of LR asymmetry in this species.

A current driven by differences in membrane potentials on the left and right side of the primitive streak is probably an upstream mechanism ultimately involved in the

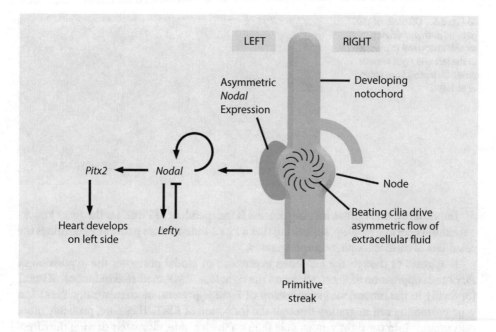

Fig. 3.4 Illustration of left-right axis formation in the mouse node. Asymmetric left-sided Nodal expression is driven by cilia to induce Lefty and Pitx2 expression. (Adapted from [7])

establishment of the left-right axis. In the mammalian node (likely also applying to the human embryo), the nodal flow ultimately expedites the leftward transport of morphogen-containing vesicles, also referred to as '*nodal vortical particles (NVP)*'. Morphogens present in the NVP include Shh and retinoic acid (RA), which possibly initiate an intracellular Ca^{2+} signalling cascade on the left side of the node, as shown in Fig. 3.4.

Alternatively, immotile cilia might sense the nodal flow via the Ca^{2+}-channel PKD2, which triggers a Ca^{2+}-wave on the left side of the node potentially leading to the upregulated expression of Nodal on the left side.

3.2 Cardiac Looping: Prrx1 and Pitx2

Prrx1 is a recently discovered homeobox transcription factor suggested to have a key role in the molecular signalling that directs cardiac looping towards the right side. In zebrafish, Ocana et al. observed a right-sided asymmetric expression domain of Prxx1a and impaired heart looping in $prxx1a^{-/-}$ mutant fish [6]. Moreover, in the chick embryo, the looping morphogenesis was impaired after knock-down of PRRX1. Interestingly, in the mouse, knockout of a Snail1 acting in the same functional context as PRXX1 also resulted in impaired looping morphogenesis.

Following cardiac looping, *Pitx2* expression is retained in the left atrium only. The knockout of *Pitx2* is associated with the development of two sinoatrial nodes (SAN) and absence of an atrial septum primum, as opposed to the physiological situation of one right-sided sinus node and a septum separating the two atria. Therefore, Pitx2 may have roles in both the suppression of SAN development on the left side and initiation of atrial septum primum formation.

3

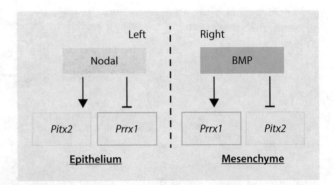

Fig. 3.5 Diagram of the genetic pathway driving asymmetric gene expression in the left and right venous poles. (Adapted from Ocana et al. [6])

Importantly, this cardiac looping process is independent of Pitx2, i.e. the *Pitx2* knockout exhibits normal looping, suggesting that a Pitx2-independent pathway determines the sidedness of cardiac looping morphogenesis.

It appears as though the left-sided expression of Nodal promotes the expression of *Pitx2* and suppression of *Prrx1*, whilst on the right side, BMP-mediated induction of Prrx1 (or Snail1 in the mouse) and suppression of *Pitx2* is present. Mechanistically, Prrx1 is a gene promoting cell migration through the induction of EMT. Therefore, probably more cells migrate into the right venous pole than on the left side, ultimately driving the direction of looping towards the right side (■ Fig. 3.5).

3.3 Development of the Cardiac Pacemaker

As discussed in ▶ Chap. 5, cardiac pacemaker potentials are rhythmically and autonomously generated by a diffusely localised collection of pacemaker cells in the intercaval sinus myocardium known as the sinoatrial node (SAN). Apart from nodal myocytes, the node consists mostly of fibroblasts embedded in an extensive extracellular matrix (ECM). Notably, SAN myocytes are smaller, with one able to distinguish three morphologically distinct cell types within the node, namely spider, spindle and extended spindle pacemaker myocytes [8].

Pacemaker myocytes have a paler appearance than working myocytes due to the presence of fewer mitochondria and a smaller number of myofibrils. Moreover, SAN myocytes are poorly electrically coupled to each other, and with the exception of some sinus exit pathways, the sinus node is electrically isolated from the surrounding myocardium by an extensive network of fibroblasts and ECM.

In the early tubular heart, the first heart field (FHF) cells all display automaticity, having the capacity to spontaneously depolarise using inward Ca^{2+} currents [9, 10]. Coupled with decelerated conduction velocities and poorly developed sarcomeres, this results in poor contractility, yet is commensurate to the oxygen and nutrient requirements of the early vertebrate embryo. The dominant site of automaticity at this stage is located on the left side of the venous pole of the heart tube, also known as the inflow tract. Importantly, it should be noted that throughout the elongation of the heart tube, the site of dominant pacemaker activity remains located on the left side of the inflow tract, suggesting that cells added to the venous pole first differentiate into pacemaker cells.

The primitive cardiac myocytes produce a single wave of electrical activation and contraction from the venous to the arterial pole of the linear heart tube, resulting in a characteristic sinusoidal electrocardiogram. During looping, most of the linear heart tube differentiates into working chamber myocardium (WCM) through the expression of Nkx2.5 and Tbx5, with Nkx2.5 suppressing Shox2 expression, a transcription factor important for SAN development. This results in the suppression of HCN4 and Tbx3, as evidenced by the absence of Nkx2.5 expression in the SAN [11].

However, a few specific regions of the heart tube actively repress the WCM gene programme in order to retain the primitive, node-like phenotype of automaticity and slow conduction. This is achieved through the Shox2-mediated inhibition of Nkx2.5, retarding the WCM gene programme. This suppression is also achieved via Tbx2 and Tbx3, two members of the T-box family of transcription factors that prevent the expression of WCM marker genes *Nppa*, and *Gja5* and *Gja1*, encoding Cx40 and Cx43, respectively. Through these mechanisms, the remnants of the primitive heart at the venous pole and atrioventricular canal continue to exhibit automaticity, supported by both the expression of Tbx3 as a marker of cardiac conduction tissue, and also the lack of Cx40 and Cx43 expression (◘ Fig. 3.6).

3.4 The SAN and AVN: Remnants of the Primitive Heart Tube

Nmyc, encoded by the *Mycn* gene, is a transcription factor associated with the proliferation of cardiac tissue, as demonstrated by the cardiac hypoplasia observed in $Mycn^{-/-}$ mice. Importantly, *Mycn* is also downregulated by Tbx2, linked with suppression of the WCM gene programme. In differentiated working chamber myocardium, Tbx20 expression suppresses Tbx2, thus antagonising *Mycn* repression and promoting cell proliferation in the WCM. Conversely, the absence of Tbx20 in the primitive pacemaking myocardium maintains Tbx2-mediated inhibition of *Mycn*, and therefore displays a low level of cell proliferation [12].

Progressively, as proliferation of the WCM continues, its tissue mass rapidly exceeds that of the primitive myocardium, meaning over time the WCM will represent the predominant phenotype of the embryonic myocardium. This change is referred to as the 'ballooning model' and is further enhanced by the addition of SHF-derived cells at both poles. Ultimately, the pacemaking tissue is progressively confined to discrete regions of the heart, i.e. the SAN and AVN (◘ Fig. 3.7).

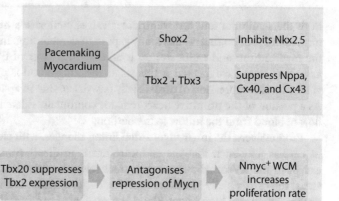

◘ **Fig. 3.6** Schematic demonstrating the transcriptional network that defines the pacemaking myocardium through suppression of the WCM gene programme

◘ **Fig. 3.7** Overview of the mechanisms that are operative in the WCM to overcome the Tbx2-mediated suppression of *MYCN* expression and cell proliferation

3

3.4.1 Tbx18

In the tubular heart, Tbx18 is initially expressed in the inflow tract, with Tbx18$^+$ mesenchymal progenitors that co-express Isl1 at the lateral-caudal border giving rise to the sinus horn and SAN. However, by the time the first morphological signs of the SAN become visible, *Tbx18* expression is confined to the right sinus horn myocardium and is co-expressed with *Isl1* and *Tbx3*. The sinoatrial node is composed of distinct head and tail regions. Interestingly, mice that lack *Tbx18* fail to form the SAN head yet retain normal pacemaking [13].

This appears to suggest that *Tbx18* has an essential role in the recruitment of mesenchymal precursors to cardiac lineage specific to the sinoatrial nodal tissue; however, it does not contribute to the functional capacity of the developing node, which is controlled by distinct transcriptional mechanisms involving Shox2, Tbx3 and Isl1.

Furthermore, Tbx18$^+$ sinus venosus precursors do not express the transcription factor Isl1 (marker of second heart field) before their differentiation, however SAN precursors co-express Isl1 and Tbx18, indicating that SAN progenitors have features of both the second heart field (Isl1$^+$) and sinus venosus progenitors (Tbx18$^+$).

3.4.2 Secondary Pacemaker

Marger et al. [14] propose that the atrioventricular node (AVN) may act as a 'secondary pacemaker', based partly upon the findings of Marionneau et al. [15], who demonstrate similar ion channel mRNA expression profiles for both the murine SAN and AVN. The latter node is secondary in nature due to the differential expression magnitudes between the two, with enhanced induction of key channels, e.g. HCN4, in the SAN. At a transcriptional level, the AVN displays expression of Tbx2 and Tbx3 competing with Tbx5 for binding to the regulatory elements of WCM genes, thus, also in case of AVN specification, repression of the WCM gene expression programme is involved.

3.5 Formation of Valves and Chambers

Both the semilunar and atrioventricular valves comprise a highly complex architecture formed in a process that retains many gaps in understanding. In the tubular heart, between the myocardium and endocardium is an extensive layer of ECM referred to as *cardiac jelly*. No valves have formed at this stage, meaning blood is propagated unidirectionally via the local compression of cardiac jelly (see discussion in ▶ Chap. 2 – the Liebau pump). As a portion of the primitive heart remains continually dilated, this ultimately guides the flow of blood from the inflow to the outflow.

Progressively, the inner myocardial lining changes, with chamber myocardium losing its cardiac jelly such that the endocardium is in direct contact with the myocardium, except in the AV canal and outflow tract regions, where cardiac jelly persists. In these areas, the cardiac jelly swells into *cardiac cushion tissue*, creating a transient solution to ensure directional blood flow. Blood is propelled at pressure through these cushions into the systemic circulation.

Eventually, endocardial cells in the atrioventricular tissue migrate into the matrix of these cushions, where they differentiate and secrete proteolytic enzymes and ECM proteins. This results in matrix remodelling, a well-studied process following initial observations by Bernanke et al. [16]. The migration of endocardial cells into the cardiac cushions is termed '*endocardial-mesenchymal transformation (EMT)*' and converts the elastic properties of the cushions to ultimately create the valvular apparatus of the adult heart. The progressive increase in valve-like ECM proteins occurs concurrently with the induction and delamination of tendon-like structures between the forming valves and myocardium, later forming the atrioventricular septa and *chordae tendinae*. The integration of the valvular apparatus with the newly formed septa, as well as a fibrous layer surrounding the valves, serves to electrically isolate the atria and ventricles.

Vital for semilunar valvulogenesis is the patterning of the great vessels, separating the common truncus into the individual architecture of the aorta and pulmonary trunk, respectively. The current consensus is that a specific population of neural crest cells (NCC) migrate from the anterior neural tube to colonise the pharyngeal arches and outflow tract, terminating migration in the outflow tract (OFT) at the level of the endocardial cushions. In a poorly understood interaction of NCC and the surrounding OFT tissues, the NCC contribute to vascular smooth muscle tissue of the aorta and pulmonary trunk, to the innervation of the heart and to the patterning of the great vessels.

3.6 Proepicardium and Coronary Artery Development

The proepicardium (PE) is an extracardiac population of mesothelial cells that develops at the base of venous inflow tract and later forms the mature epicardium. All epicardium-derived cells (EPDCs), including cardiac fibroblasts and the coronary vasculature, derive from the proepicardium. Prior to this, the transmural wall only comprises the endocardium and myocardium.

It is important to discuss epicardial development in both the chick and murine embryo, as these constitute two well-investigated yet distinct processes.

3.6.1 Chick Embryo

The PE originates as an asymmetrical outgrowth of mesothelial cells on the right sinus horn that expands over time. A small cluster of PE cells is also formed on the left side, however is lost through apoptosis. Eventually, villous projections from the PE protrude towards the inner curvature of the heart tube [17]. Contact between the myocardial surface and the PE projection forms a functional tissue bridge, through which PE cells migrate and colonise the myocardial surface, forming the epicardium. This colonisation process also involves an extracellular-rich sub-epicardial mesenchyme layer forming underneath the epicardium.

3.6.2 Murine Embryo

The PE originates from a similar embryonic location as in the chick, however no tissue bridge is formed in order to establish contact with the ventricular wall surface. Instead, proepicardial cells are released as free-floating vesicles that establish epicardial islands on the myocardium, facilitating colonisation of the wall. This notion was challenged by

3

Rodgers et al. however, who observed that whilst free-floating vesicles are present, proepicardial projections do directly attach to the heart [18]. This gave rise to a revised common view that murine PE projections directly attach to the heart, however in a transient manner. The surface of the proepicardium retracts rapidly from the myocardium, but the transient contact is sufficient for the direct transfer of PE vesicles, which together with free-floating vesicles, establishes the colonisation of the mouse heart with PE-derived cells.

3.6.3   Marker Genes to Consider

There are multiple key signalling molecules to consider in this process [19]. Two evolutionarily conserved marker genes of the forming PE are WT1 and Tbx18, both of which are strongly associated with PE development and the latter also with the development of the venous pole. Nkx2.5 has been demonstrated as instrumental for normal PE development, with the knockout exhibiting a severe reduction in PE cells.

Importantly though, it should be mentioned that due to the general importance of Nkx2.5 in cardiac development, knockout of this gene may have extensive secondary effects and therefore may indirectly impact PE development. Equally important is Gata4, which is expressed in the PE in a BMP-dependent manner and whose knockout results in a complete loss of PE formation [19].

Finally, the epicardium orchestrates the growth of the myocardial wall at midgestation through the secretion of paracrine signals. The myocardial wall consists of an outer *compact layer*, which is responsible for the growth of the chamber wall, and an inner layer of differentiated cells proliferating at lower levels, termed the *trabecular layer*, which is mostly responsible for the contractile work of the ventricular wall at this stage of development.

Although the signalling nexus for this pathway is poorly understood, molecular data of the murine and chicken embryo identifies a significant role for epicardium-derived *retinoic acid* and *erythropoietin* (*EPO*), both of which are paracrine signals involved in promoting the proliferation of cells in the compact layer myocardium, potentially acting indirectly (non-cell autonomously), through the stimulation of secretion of fibroblast growth factors (FGFs) from the epicardium. The importance of these signals becomes apparent following inhibition of epicardialisation, resulting in a thin compact layer myocardium, which is lethal due to embryonic heart failure.

3.7   What We Don't Know: Potential Presence of a Tertiary Heart Field

Based on what has already been discussed in this chapter, it would appear that cardiac pacemaking tissue exists as isolated remnants of the primitive tubular heart myocardium that were prevented from entering WCM differentiation through suppression of the working chamber myocardium gene programme via the inhibitory activity of Tbx2, Tbx3, Tbx18 and Shox2 [10].

However, an alternative proposition of the origin of cardiac pacemaker tissue exists, based upon a study in the chicken embryo [20]. Bressan et al. demonstrated the origin of the SAN myocardium from mesoderm, which at gastrula stage is located in a significantly more posterior location and is therefore distinct from the mesoderm population that gives rise the first and second heart fields identifiable by the expression domains of Nkx2.5 and Isl-1, respectively [19].

Wnt signalling, in addition to other yet unidentified signalling factors, induce pacemaker-like cells from this so-called 'tertiary heart field'.

These cells enter the heart only at later stages, during heart looping. While the heart is already beating for more than a day before the definitive pacemaker lineage reach the heart, it is this lineage that constitutes the main source of pacemaker cells in the chicken SAN.

The competing theories of the origin of pacemaker cells are established by carefully executed experiments. Genetic experiments in the mouse and embryological experiments in the chick both establish the origin of the SAN. The model derived from the investigation of murine embryos suggests a common origin of pacemaker cells and chamber myocytes. Through the expression of Tbx3 and other inhibitory transcription factors, putative pacemaker cells are hindered from becoming working myocytes. In contrast, the model derived from work in the chicken embryo suggests an exclusive origin of pacemaker cell from a lineage different from the first and second heart fields. A weakness of the tertiary heart field model is its inability to explain both the molecular similarity of the SAN and AVN, and the developmental origin of the AV node.

Clearly, as the models rely on different species, further research is needed to comparatively investigate the origin of the pacemaker tissue, in particular focussing on the whether a tertiary heart field is also present in mammals. Moreover, it is possible that the SAN may receive cellular input from a number of cell lineages with different developmental origins; for example, formation of the SAN head is dependent on Tbx18, while the tail is Tbx18-independent [13]. Further insight into this question may have important implications, both in our understanding of sinus node development, but also for the generation of biological pacemakers.

3.8 Where We're Heading: Ectopic Tbx18 Induction

The investigation of transcription factors involved in sinus node development has prompted investigations into their capacity to convert cardiac myocytes into pacemaker cells [21]. From these studies, Tbx18 has proved to be the most potent transcription factor to convert neonatal rat cardiac myocytes into pacemaker cells exhibiting automaticity [21, 22].

Subsequent to this, the adenoviral gene transfer of Tbx18 into the porcine heart was sufficient to transiently induce a biological pacemaker, thereby overcoming surgically induced complete heart block [23].

Notwithstanding these encouraging findings, the transgenic expression of *Tbx18* into working cardiomyocytes of the embryonic mouse heart was unable to induce an SAN-like phenotype, however it did lead to pathological structural changes, altered gene expression and ultimately resulted in neonatal lethality [24]. Clearly, further research is required to fully investigate the range of potential side effects that may accompany the ectopic induction of an SAN gene programme with the help of Tbx18.

3

Take-Home Message

- In the chicken embryo, both asymmetric gene expression and the presence of cilia at Hensen's node contribute to the establishment of the three main embryonic axes.
- Cardiac looping has been extensively studied in the zebrafish, where it has been identified as independent of Pitx2, suggesting a Pitx2-independent pathway determines the sidedness of cardiac looping morphogenesis.
- A study in the chicken embryo has identified the origin of the SAN myocardium from a mesoderm population located posterior to those that give rise to the first and second heart fields.
- Studies in neonatal rat cardiac myocytes have identified Tbx18 as a potent transcription factor for the induction of automaticity in cardiac myocytes, however technical challenges persist.

References

1. Brand T (2003) Heart development: molecular insights into cardiac specification and early morphogenesis. Dev Biol 258:1–19
2. Monsoro-Burq A, Le Douarin NM (2001) BMP4 plays a key role in left-right patterning in chick embryos by maintaining Sonic Hedgehog asymmetry. Mol Cell 7:789–799
3. Campione M et al (2001) Pitx2 expression defines a left cardiac lineage of cells: evidence for atrial and ventricular molecular isomerism in the iv/iv mice. Dev Biol 231:252–264
4. Gros J, Feistel K, Viebahn C, Blum M, Tabin CJ (2009) Cell movements at Hensen's node establish left/right asymmetric gene expression in the chick. Science 324:941–944
5. Nonaka S, Tanaka Y, Okada Y, Takeda S, Harada S, Kanai Y, Kido M, Hirokawa N (1998) Randomization of left-right asymmetry due to loss of nodal cilia generating leftward flow of extraembryonic fluid in mice lacking KIF3B motor protein. Cell 95:829–837
6. Ocana OH et al (2017) A right-handed signalling pathway drives heart looping in vertebrates. Nature 549:86–90
7. Alberts B, Johnson A, Lewis J, Raff M, Roberts K, Walter P (2008) Molecular biology of the cell. Garland Science, New York. Figure 22-87a
8. Masson-Pevet MA et al (1984) Pacemaker cell types in the rabbit sinus node; a correlative ultrastructural and electrophysiological study. J Mol Cell Cardiol 16:53–63
9. Lakatta EG, Maltsev VA, Vinogradova TM (2010) A coupled SYSTEM of intracellular Ca2+ clocks and surface membrane voltage clocks controls the timekeeping mechanism of the heart's pacemaker. Circ Res 106:659–673
10. DiFrancesco D (2010) The role of the funny current in pacemaker activity. Circ Res 106:434–444
11. Van Weerd JH, Christoffels VM (2016) The formation and function of the cardiac conduction system. Development 53:197–210
12. Cai X et al (2011) Myocardial Tbx20 regulates early atrioventricular canal formation and endocardial epithelial-mesenchymal transition via Bmp2. Dev Biol 360:381–390
13. Wiese C, Grieskamp T, Airik R, Mommersteg MT, Gardiwal A, De Gier-de Vries C et al (2009) Formation of the sinus node head and differentiation of sinus node myocardium are independently regulated by Tbx18 and Tbx3. Circ Res 104:388–397
14. Marger L et al (2011) Pacemaker activity and ionic currents in mouse atrioventricular node cells. Channels (Austin) 5:241–250
15. Marionneau C, Couette B, Liu J, Li H, Mangoni ME, Nargeot J, Lei M, Escande D, Demolombe S (2005) Specific pattern of ionic channel gene expression associated with pacemaker activity in the mouse heart. J Physiol 562:223–234
16. Bernanke DH, Markwald RR (1982) Migratory behaviour of cardiac cushion tissue cells in a collagen-lattice culture system. Dev Biol 91:235–245

17. Männer J (1992) The development of pericardial villi in the chick embryo. Anat Embryol 186:379–385
18. Rodgers LS, Lalani S, Runyan RB, Camenisch TD (2008) Differential growth and multicellular villi direct proepicardial translocation to the developing mouse heart. Dev Dyn 237:145–152
19. Schlueter J, Brand T (2011) Origin and fates of the proepicardium. Aswan Heart Centre Science & Practice Series. QScience, Hamad Bin Khalifa University Press, Doha, Quatar, p 11
20. Bressan M, Liu G, Mikawa T (2013) Early mesodermal cues assign avian cardiac pacemaker fate potential in a tertiary heart field. Science 340:744–748
21. Brand T (2016) Tbx18 and the generation of a biological pacemaker. Are we there yet? J Mol Cell Cardiol 97:263–265
22. Kapoor N, Galang G, Marban E, Cho HC (2011) Transcriptional suppression of connnexin43 by TBX18 undermines cell-cell electrical coupling in postnatal cardiomyocytes. J Biol Chem 286:14073–14079
23. Hu YF, Dawkins JF, Cho HC, Marban E, Cingolani E (2014) Biological pacemaker created by minimally invasive somatic reprogramming in pigs with complete heart block. Sci Transl Med 6:245ra94
24. Greulich F, Trowe MO, Leffler A, Stoetzer C, Farin HF, Kispert A (2016) Misexpression of Tbx18 in cardiac chambers of fetal mice interferes with chamber-specific developmental programs but does not induce a pacemaker-like gene signature. J Mol Cell Cardiol 97:140–149

Myocardial Microstructure and Contractile Apparatus

Umamah Tarvala and Zohya Khalique

© Springer Nature Switzerland AG 2019
C. Terracciano, S. Guymer (eds.), *Heart of the Matter*, Learning Materials in Biosciences,
https://doi.org/10.1007/978-3-030-24219-0_4

What You Will Learn in This Chapter

This chapter will address the anatomy of the heart from a microstructural perspective and review how the organisation of cardiomyocytes supports cardiac function. Both the helical arrangement of cardiomyocytes and the orientation of sheetlets will be examined. The potential of a novel technique called diffusion tensor cardiovascular magnetic resonance (DT-CMR) will be discussed, in particular its ability to non-invasively assess the cardiac microstructure in both health and disease.

4

Learning Objectives
- Acknowledge the complexity of the myocardial microstructure.
- Understand how the configuration of cardiomyocytes and sheetlets relate to ventricular contractile function.
- Appreciate how novel non-invasive techniques will aid our understanding of the dynamics of the myocardial microstructure.

4.1   Cardiac Microstructure

The cardiac microstructure is complex, and its dynamics are not yet fully understood. It is often considered to have a hierarchical structure. Sarcomeres are the basic contractile units of the myocardium and are aligned in series to form myofibrils, which themselves assemble within the cardiomyocytes (CMs) in parallel. CMs connect end to end with several others in a branching configuration [1]. Moreover, CMs aggregate into secondary structures called *sheetlets* that underlie the laminar nature of the myocardium [2, 3]. There is an important relationship between the microstructure and overall cardiac function.

4.1.1   Sarcomeres

Sarcomeres are formed of thin and thick filaments and are approximately 2 μm long. They are bounded by Z-discs [4]. The thin filament comprises actin molecules associated with the troponin complex (troponins T, I and C) and tropomyosin. Tropomyosin rests over the myosin-binding site of the actin filament [5]. The thick filament comprises two myosin heavy chains and four myosin light chains. In each thick filament, there are three key regions: the motor head, the hinge at the neck and the tail.

The arrangement of the filaments gives rise to alternating light (I) and dark (A) bands [4]. The lighter I bands contain only the thin filament, whilst the darker A band includes the thick filament as well as the area of overlap with the thin filament. The A band also contains the H zone, a region of no actin. The M line reflects the centre of the sarcomere, with the protein *titin* running from the M line to the Z disc.

4.1.2   Cardiomyocytes

CMs are multinucleated cells, with a principal long axis and multiple intercalations with other CMs. There is a surprising lack of agreement in the terminology relating to CM organisation, with the term 'fibre' being used to describe both individual CMs and groups of CMs [6]. Others avoid the term altogether to retain the distinction between cardiac and skeletal muscle.

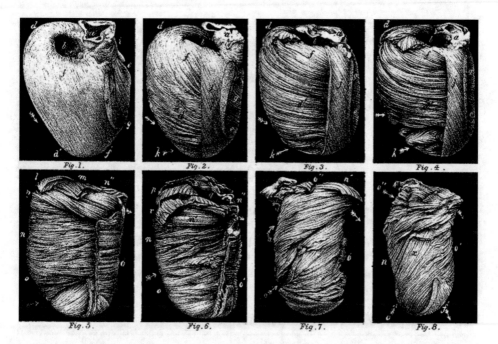

Fig. 4.1 The left ventricle (LV) has a helical arrangement. The LV is dissected from the epicardium (top left) through several layers to the endocardium (bottom right). There is transmural variation in the orientation of the cardiomyocytes. (Image from [7])

Gross dissection of mammalian hearts reveals a 'grain-like' appearance to the myocardium, reflecting the overall direction of CM long axis orientation [7]. The CMs have a helical orientation, as shown in ◼ Fig. 4.1. In the epicardium, CMs are orientated in a negative left-handed (LH) helix, rotated through to a circumferential alignment in the mesocardium and progress to a positive right-handed (RH) helix in the endocardium. Both light [8] and extended volume confocal microscopy [9] techniques confirm this transmural variation of the cardiomyocyte orientation from approximately −70° at the epicardium to +70° at the endocardium (where the angle lies between the CM long axis and the circumferential direction and is projected onto the plane tangential to the epicardium).

4.1.3 Sheetlets

A secondary level of CM organisation exists, in which groups of approximately 8–12 CMs aggregate, surrounded by collagenous perimysium and separated by shear layers [2, 3]. These grouped CMs are termed sheetlets, shown in ◼ Fig. 4.2. Originally, it was believed that sheetlets spanned the entire width of the LV wall; however, it is now appreciated that there are multiple sub-populations of sheetlets with differing orientations throughout the LV [11–13].

An important collagenous network exists, interlaced with the CMs and sheetlets. In the subepicardium, long interconnected cords of collagen run parallel to the long axis of CMs. In the mid-wall, perimysial collagen surrounds sheetlets, forming part of the cleavage planes that are seen in ◼ Fig. 4.2. Longitudinal cords (base-apex) are less frequently

4

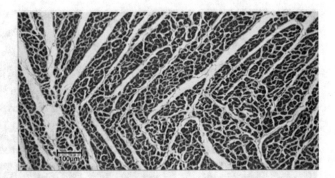

◻ **Fig. 4.2** Histology of porcine myocardium from the mid-left ventricular wall. Cardiomyocytes are seen in cross section and aggregate into groups, separated by the white fissures or 'shear layers'. (Adapted from [10])

observed in the mesocardium. In the subendocardium, there is a dominance of the cleavage plane structure in the collagen network, suggesting that the sheetlet configuration is increasingly well-defined towards the endocardium [2]. These cleavage planes are not simply space between sheetlets or distortion artefacts from histological preparation, but in fact play an important role in the dynamic nature of the myocardium.

4.2 Structure–Function Relationship

Cardiac function is complex, with multidirectional myocardial deformations occurring across the cardiac cycle. The interplay of radial thickening and thinning, longitudinal and circumferential shortening, lengthening, and torsion all contribute to LV function. Whilst ejection fraction (EF) is the most ubiquitous of parameters used to define contractile function, it measures luminal change, rather than myocardial mechanics directly. *Strain* is a measure of myocardial deformation (relative change from resting diastolic dimension) and is typically described by dimensionless units. Global longitudinal strain has been shown to be superior to EF in predicting major adverse cardiac events [14].

Radial strain reflects LV wall thickening and is positive, typically in the range of 0.40. In contrast, circumferential and longitudinal strain are both negative, and approximately -0.20 [15, 16]. These strains are in the direction of the normal orthogonal directions, but more complex cross-directional deformations known as shear also exist.

4.2.1 Sarcomeric Function

These various strains are underpinned by the contractile force generated by the sarcomere. It is the dynamic interaction of the thick and thin filament that produces contractile force. This is known as the '*sliding filament theory*' [4, 17]. CM membrane depolarisation results in the release of Ca^{2+} from the sarcoplasmic reticulum, which binds to troponin C. This, as well as phosphorylation of troponin I results in tropomyosin rotating around the actin filament and opening the myosin binding site [5, 18]. Consequently, the myosin head is able to form a crossbridge with the actin filament and slide it along towards the centre of the A band in a 'power stroke'. Adenosine triphosphate (ATP) then binds to the myosin head, and the crossbridge detaches.

Sarcomere length reduces in systole to approximately 80–65% of its diastolic length, due to this sliding filament motion. It is recognised that there is a length–tension relationship of increasing force generation as diastolic sarcomere length increases. This accords with the ascending limb of the Frank-Starling law curve; increasing contractility (cardiac output) as diastolic filling (end-diastolic volume) rises. This will be explained further in ▶ Chap. 10 [19].

4.2.2 Relating Cardiomyocytes to Cardiac Function

There is an intuitive relationship between sarcomeric shortening of ~15% and CM shortening of a similar magnitude. Indeed, this value approximates to the negative circumferential strain seen in the mid-wall where CMs are circumferentially aligned [17, 20]. What is less clear is how the associated ~8% CM thickening translates into wall thickening of 40% – a paradox addressed by many groups over recent decades [8, 21–23].

Initially, it was recognised that cleavage plane orientation differed in different contractile states, supporting the concept of redistribution of myocardial volume through the cardiac cycle [21]. The identification of sheetlets, married with experiments measuring shear strains, developed the subsequent explanation of sheetlets sliding relative to one another, facilitating wall thickening beyond individual CM thickening [22–24]. However, such shear analyses do not directly assess sheetlet orientation, with these studies further limited by the static nature of histology.

Recent novel developments in a technique called *diffusion tensor cardiovascular magnetic resonance* (DT-CMR) have been instrumental in developing our understanding of this conundrum [25]. DT-CMR is a unique non-invasive tool that can evaluate the cardiac microstructure in vivo by probing the diffusion of water through the myocardium. Both the mean CM long axis orientation and the dominant sheetlet direction can be obtained, as demonstrated in ◧ Fig. 4.3 [26–28]. The technique has been histologically validated through various experiments, including in vivo, in situ, ex vivo and histological correlation [26, 28–31]. It has been applied in humans to give dynamic information about the microstructural rearrangements that are integral to wall thickening and LV contraction [26]. In diastole, sheetlets are orientated parallel to the LV epicardial wall; however, in systole they reorient to a more wall-perpendicular alignment, facilitated by the shear layers. This degree of reorientation is referred to as *sheetlet mobility*.

Interestingly, the helix angle does not undergo substantial change between diastole and systole, instead retaining the negative-to-positive helix angle progression that appears to be preserved throughout the mammalian species [7, 26]. Rather, the helical arrangement is key in driving myocardial rotation and torsion [32]. CM shortening of the LH epicardial myocytes leads to clockwise rotation basally and anticlockwise rotation apically, when viewed from the apex. Contraction of the RH subendocardial myocytes results in anticlockwise rotation basally and clockwise rotation apically. There is a net clockwise rotation basally and anticlockwise rotation apically, due to the dominant effect of the larger epicardial radius [33].

These opposing rotations give rise to LV torsion – the wringing effect that aids effective LV emptying. The consequent recoil in diastole facilitates passive LV filling [34]. Hence, there is an important relationship between the helical arrangement of cardiomyocytes and LV function.

4

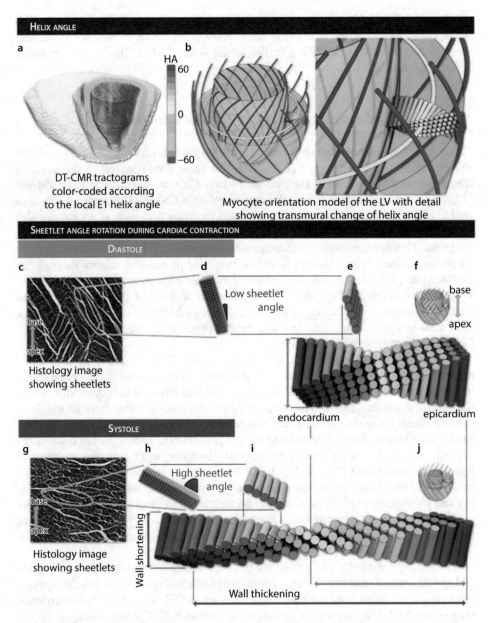

☐ **Fig. 4.3** DT-CMR provides cardiomyocyte and sheetlet orientation. In the top row, helix angle (HA) is depicted. DT-CMR tractography demonstrates the known transmural variation from a negative HA at the epicardium (blue) through to a positive HA in the endocardium (red). In the bottom panel, the sheetlet angle is demonstrated with porcine histology alongside. In diastole, histology (c) shows a more wall-parallel sheetlet orientation, which is a low sheetlet angle (d–f). In systole, a more wall-perpendicular orientation occurs, shown in histology (g), which translates to a higher sheetlet angle (h, I). While cardiomyocyte orientation remains similar in both cardiac phases, there is a significant increment in sheetlet angle, which marries with the increased wall thickness (j). (Image from [26])

4.3 Clinical Implications: DT-CMR in Microstructural Derangement

DT-CMR is the only tool that allows for the non-invasive in vivo characterisation of the myocardial microstructure. It has offered novel insight into our understanding of myocardial mechanics in health, but also in disease. In acute myocardial infarction, the proportion of RH helical structures decreases, concordant with the susceptibility of the endocardium to ischaemia. Subsequent to this, a gain of LH cardiomyocytes in the infarct zone and RH cardiomyocytes in the remote zone are thought to be adaptive remodelling responses [35, 36]. Disturbance of the microarchitecture post-infarct has been characterised by DT-CMR, which may offer a contrast-free method of delineating cardiac fibrosis, in addition to identifying a potential new tool to guide catheter ablation procedures for infarct-driven ventricular arrhythmias [37].

DT-CMR has also identified novel abnormalities in sheetlet behaviour in cardiomyopathy. In hypertrophic cardiomyopathy (HCM), sheetlets have been demonstrated as retaining a more systolic-like orientation (more wall-perpendicular) in diastole, with reduced sheetlet mobility [10, 26]. The failure of sheetlets to return to a more wall-parallel orientation in diastole can be described as a 'failure of diastolic relaxation'. Interestingly, this microstructural abnormality builds upon our knowledge of HCM pathophysiology. Many sarcomeric mutations result in increased myofilament sensitivity to calcium, increasing relative CM tension and contributing to the diastolic dysfunction in HCM [38, 39]. In contrast, sheetlets in dilated cardiomyopathy (DCM) have been observed as failing to re-align to the expected systolic wall-perpendicular angulation, instead retaining a more diastolic-like configuration [26].

In both of these cardiomyopathies, radial strain is similarly reduced; however, DT-CMR is able to identify the respective sheetlet abnormalities, providing a deeper understanding of the pathophysiology underlying the diseases. HCM and other forms of cardiomyopathy will be expanded upon in greater detail in ► Chap. 12.

In congenital heart disease, DT-CMR of patients with situs inversus totalis has disproven the idea that the condition is one of simply gross positional abnormality. Instead, gross derangement of the helical arrangement was demonstrated, with an inverted pattern basally that transitions to a more solitus pattern apically (◘ Fig. 4.4). In the mid-ventricular transition zone, decreased sheetlet mobility and strain were identified, whilst overall absolute torsion was reduced [40].

4.4 Where We're Heading

Our understanding of the cardiac microstructure and its relationship with cardiac function continues to grow. Novel tools such as DT-CMR are critical to developing our appreciation of this complex interplay. Whilst the sheetlets and their dynamics during the cardiac cycle can be demonstrated in vivo, it is unclear whether wall thickening results from microstructural rearrangement or vice versa. In the same vein, does impaired sheetlet mobility reflect or drive the cardiomyopathic process? There are inherent differences in the morphology of the right and left ventricles and so the microstructural assessment of both ventricles needs further study.

4

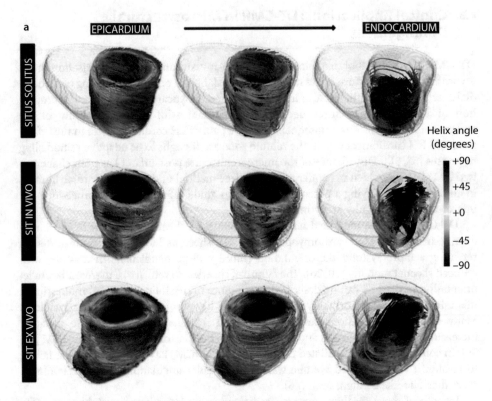

☐ **Fig. 4.4** CM orientation is deranged in situs inversus. DT-CMR demonstrates the expected helical arrangement in situs solitus, whereby CM orientation progresses from a negative helix (blue) at the epicardium towards a positive helix angle (red) at the endocardium. Neither the in-vivo and ex-vivo situs inversus hearts exhibit this profile and instead show inversion at the base that trends towards the situs solitus arrangement at the apex. There is a mid-ventricular transition zone. Adapted from [40]

Take-Home Message

- The myocardial microstructure is complex, with multiple levels of organisation. Cardiac function is more than the ejection fraction. Various myocardial deformations take place through the cardiac cycle, including radial thickening, circumferential and longitudinal shortening, as well as torsion.
- Cardiomyocytes have a helical arrangement that transitions from left-handed in the epicardium through circumferential in the mesocardium and right-handed in the endocardium. This organisation is typically preserved in mammals and drives cardiac torsion.
- A second level of cardiomyocyte organisation exists, in which aggregated cardiomyocytes form sheetlets. These sheetlets re-orientate through the cardiac cycle and are integral to the process of wall thickening.
- DT-CMR is a novel tool that can non-invasively offer information about cardiomyocyte and sheetlet orientations in vivo. It has identified sheetlet abnormalities in cardiomyopathy, deranged cardiomyocyte structure in congenital heart disease and can assess microarchitectural changes following myocardial infarction.

References

1. Braunwald E, Ross J, Sonnenblick EH (1967) Mechanisms of contraction of the normal and failing heart. N Engl J Med 277:910
2. Pope AJ, Sands GB, Smaill BH, LeGrice IJ (2008) Three-dimensional transmural organization of perimysial collagen in the heart. AJP Hear Circ Physiol. 295:H1243
3. LeGrice I, Smaill B, Chai L, Edgar S, Gavin J, Hunter P (1995) Laminar structure of the heart: ventricular myocyte arrangement and connective tissue architecture in the dog. Am J Physiol Heart Circ Physiol 269:H571
4. Huxley AF, Niedergerke R (1954) Structural changes in muscle during contraction: interference microscopy of living muscle fibres. Nature 173(4412):971–973
5. Lehman W, Craig R, Vibert P (1994) Ca2+-induced tropomyosin movement in Limulus thin filaments revealed by three-dimensional reconstruction. Nature 368(6466):65–67
6. Gilbert SH, Benson AP, Li P, Holden AV (2007) Regional localisation of left ventricular sheet structure: integration with current models of cardiac fibre, sheet and band structure. Eur J Cardiothorac Surg 32:231
7. Pettigrew JB (1864) On the arrangement of the muscular fibres in the ventricles of the vertebrate heart, with physiological remarks. Philos Trans R Soc Lond 154:445–500
8. Streeter DD, Spotnitz HM, Patel DP, Ross J, Sonnenblick EH (1969) Fiber orientation in the canine left ventricle during diastole and systole. Circ Res 24(3):339–347
9. Sands GB, Smaill BH, LeGrice IJ (2008) Virtual sectioning of cardiac tissue relative to fiber orientation. In: 2008 30th annual international conference of the IEEE engineering in medicine and biology society
10. Ferreira PF, Kilner PJ, Mcgill LA, Nielles-Vallespin S, Scott AD, Ho SY et al (2014) In vivo cardiovascular magnetic resonance diffusion tensor imaging shows evidence of abnormal myocardial laminar orientations and mobility in hypertrophic cardiomyopathy. J Cardiovasc Magn Reson 16:87
11. Helm PA, Younes L, Beg MF, Ennis DB, Leclercq C, Faris OP et al (2006) Evidence of structural remodeling in the dyssynchronous failing heart. Circ Res 98:125
12. Kung GL, Nguyen TC, Itoh A, Skare S, Ingels NB, Miller DC et al (2011) The presence of two local myocardial sheet populations confirmed by diffusion tensor MRI and histological validation. J Magn Reson Imaging 34:1080
13. Harrington KB (2005) Direct measurement of transmural laminar architecture in the anterolateral wall of the ovine left ventricle: new implications for wall thickening mechanics. AJP Hear Circ Physiol. 288(3):H1324–H1330
14. Kalam K, Otahal P, Marwick TH (2014) Prognostic implications of global LV dysfunction: a systematic review and meta-analysis of global longitudinal strain and ejection fraction. Heart 100:1673
15. Moore CC, Lugo-Olivieri CH, McVeigh ER, Zerhouni EA (2000) Three-dimensional systolic strain patterns in the normal human left ventricle: characterization with tagged MR imaging. Radiology 214:453
16. Yingchoncharoen T, Agarwal S, Popović ZB, Marwick TH (2013) Normal ranges of left ventricular strain: a meta-analysis. J Am Soc Echocardiogr 26:185
17. Huxley H, Hanson J (1954) Changes in the cross-striations of muscle during contraction and stretch and their structural interpretation. Nature 173(4412):973–976
18. Layland J, Solaro RJ, Shah AM (2005) Regulation of cardiac contractile function by troponin I phosphorylation. Cardiovasc Res 66:12
19. Fabiato A, Fabiato F (1975) Dependence of the contractile activation of skinned cardiac cells on the sarcomere length. Nature 256:54
20. Sonnenblick EH, Ross J, Covell JW, Spotnitz HM, Spiro D (1967) The ultrastructure of the heart in systole and diastole. Chantes in sarcomere length. Circ Res 21:423
21. Spotnitz HM, Spotnitz WD, Cottrell TS, Spiro D, Sonnenblick EH (1974) Cellular basis for volume related wall thickness changes in the rat left ventricle. J Mol Cell Cardiol 6:317
22. LeGrice IJ, Takayama Y, Covell JW (1995) Transverse shear along myocardial cleavage planes provides a mechanism for normal systolic wall thickening. Circ Res 77:182
23. Cheng A, Nguyen TC, Malinowski M, Daughters GT, Miller DC, Ingels NB (2008) Heterogeneity of left ventricular wall thickening mechanisms. Circulation 118:713
24. Costa KD, Takayama Y, McCulloch AD, Covell JW (1999) Laminar fiber architecture and three-dimensional systolic mechanics in canine ventricular myocardium. Am J Physiol Circ Physiol. 276:H595

4

25. Nielles-Vallespin S, Mekkaoui C, Gatehouse P, Reese TG, Keegan J, Ferreira PF et al (2013) In vivo diffusion tensor MRI of the human heart: reproducibility of breath-hold and navigator-based approaches. Magn Reson Med 70:454
26. Nielles-Vallespin S, Khalique Z, Ferreira PF, de Silva R, Scott AD, Kilner P et al (2017) Assessment of myocardial microstructural dynamics by in vivo diffusion tensor cardiac magnetic resonance. J Am Coll Cardiol 69:661
27. Reese TG, Weisskoff RM, Smith RN, Rosen BR, Dinsmore RE, Wedeen VJ (1995) Imaging myocardial fiber architecture in vivo with magnetic resonance. Magn Reson Med 34:786
28. Scollan DF, Holmes A, Winslow R, Forder J (1998) Histological validation of myocardial microstructure obtained from diffusion tensor magnetic resonance imaging. Am J Physiol Circ Physiol 275:H2308
29. Holmes AA, Scollan DF, Winslow RL (2000) Direct histological validation of diffusion tensor MRI in formaldehyde- fixed myocardium. Magn Reson Med 44:157
30. Hsu EW, Muzikant AL, Matulevicius SA, Penland RC, Henriquez CS (1998) Magnetic resonance myocardial fiber-orientation mapping with direct histological correlation. Am J Physiol Circ Physiol. 274:H1627
31. Chen J (2005) Regional ventricular wall thickening reflects changes in cardiac fiber and sheet structure during contraction: quantification with diffusion tensor MRI. AJP Hear Circ Physiol 289:H1898
32. Rüssel IK, Götte MJW, Bronzwaer JG, Knaapen P, Paulus WJ, van Rossum AC (2009) Left ventricular torsion. An expanding role in the analysis of myocardial dysfunction. JACC Cardiovasc Imaging 2(5):648–655
33. Sengupta PP, Tajik AJ, Chandrasekaran K, Khandheria BK (2008) Twist mechanics of the left ventricle. Principles and application. JACC Cardiovasc Imaging 1(3):366–376
34. Young AA, Cowan BR (2012) Evaluation of left ventricular torsion by cardiovascular magnetic resonance. J Cardiovasc Magn Reson 14:49
35. Wu MT, Su MY, Huang YL, Chiou KR, Yang P, Pan HB et al (2009) Sequential changes of myocardial microstructure in patients postmyocardial infarction by diffusion-tensor cardiac mr correlation with left ventricular structure and function. Circ Cardiovasc Imaging 2:32
36. Wu MT, Tseng WYI, Su MYM, Liu CP, Chiou KR, Wedeen VJ et al (2006) Diffusion tensor magnetic resonance imaging mapping the fiber architecture remodeling in human myocardium after infarction: correlation with viability and wall motion. Circulation 114:1036
37. Mekkaoui C, Jackowski MP, Kostis WJ, Stoeck CT, Thiagalingam A, Reese TG et al (2018) Myocardial scar delineation using diffusion tensor magnetic resonance tractography. J Am Heart Assoc 7(3):e007834
38. Pinto JR, Parvatiyar MS, Jones MA, Liang J, Ackerman MJ, Potter JD (2009) A functional and structural study of troponin C mutations related to hypertrophic cardiomyopathy. J Biol Chem 284:19090
39. Willott RH, Gomes AV, Chang AN, Parvatiyar MS, Pinto JR, Potter JD (2010) Mutations in Troponin that cause HCM, DCM AND RCM: what can we learn about thin filament function? J Mol Cell Cardiol 48:882
40. Khalique Z, Ferreira PF, Scott AD, Nielles-Vallespin S, Kilner PJ, Kutys R et al (2018) Deranged myocyte microstructure in situs inversus totalis demonstrated by diffusion tensor cardiac magnetic resonance. JACC Cardiovasc Imaging 11(9):1360–1362

An Introduction to the Cardiac Action Potentials

Elizabeth Thong, Ayesha Ahmed, and Kenneth T. MacLeod

© Springer Nature Switzerland AG 2019
C. Terracciano, S. Guymer (eds.), *Heart of the Matter*, Learning Materials in Biosciences,
https://doi.org/10.1007/978-3-030-24219-0_5

What You Will Learn in This Chapter

This chapter will begin by examining the origin of the ventricular action potential (AP), its constituent ionic currents and the channels through which these currents flow. Several examples will be used to help appreciate the clinical implications of modifying this excitation system. Finally, we will introduce the pacemaker potential and the contemporary theories surrounding its generation.

Learning Objectives

— Analyse the five phases of the cardiac AP, state the main ionic currents active in each phase and the channels through which these currents flow.
— Provide three examples of the clinical relevance of the aforementioned channels.
— Consider the contemporary theories of pacemaker potential generation.

5.1 Introducing the Cardiac Action Potential

The cardiac AP provides the electrical component of excitation-contraction coupling, using voltage changes across the cell membranes. The length of the AP plays a role in determining the strength of cardiomyocyte contraction. To ensure effective pumping function, cardiac muscle requires a long, slow contraction, and thus the ventricular AP is relatively long (~350 ms) compared with the excitatory events characteristic of the peripheral nervous system (~2 to 3 ms). The shape and duration of the AP varies in different parts of the heart largely determined by the size and type of ionic currents that flow. In turn, ionic currents are regulated by the levels of expression of ion channels through which they flow [2]. The initial cardiac AP (termed the 'pacemaker potential') is generated both autonomously and rhythmically by the cells of the *sinoatrial node (SAN)*: a diffuse collection of specialised pacemaker cells located around the border between the superior vena cava and the right atrium [1]. These cells have different characteristics and so, we will discuss the pacemaker potential separately from the ventricular AP.

5.2 The Ventricular Action Potential

The ventricular AP can be considered as the summation of five phases, each of which possesses a unique ionic profile based on the transient activation of ion transporters. This creates the broad morphology shown in ◘ Fig. 5.1 [3]. The five phases are:
— $\phi\,0$ – Upstroke
— $\phi\,1$ – Early Repolarisation
— $\phi\,2$ – Plateau
— $\phi\,3$ – Final Repolarisation
— $\phi\,4$ – Resting Membrane Potential (RMP)

5.2.1 $\phi\,0$ – Upstroke

The arrival of an action potential (AP) from a neighbouring cell causes a small depolarisation of the cardiomyocyte. If this depolarisation is sufficient to bring the membrane potential to a threshold level sensed by a large number of voltage-gated Na^+ channels,

Fig. 5.1 A typical AP from a ventricular myocyte showing phases and the approximate lengths of the absolute (ARP) and relative (RRP) refractory periods . The main changes in the relative permeability of the myocyte membrane to ions carried by the main currents during each phase are shown

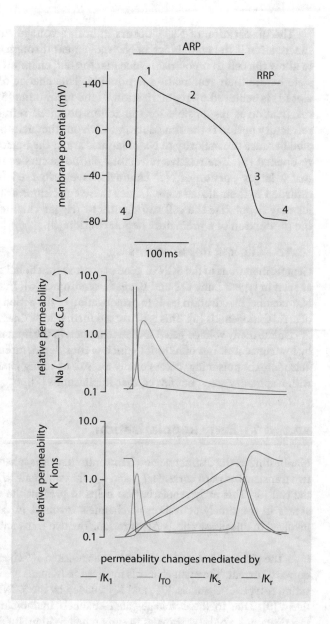

these open with the subsequent influx of Na^+ ions leading to a rapid depolarisation of the cell towards the Na^+ equilibrium potential [4]. This is known as the upstroke phase.

The inward Na^+ current is usually abbreviated as I_{Na} and influx of ions occurs via the $Na_{v1.5}$ channel, encoded by the *SCN5A* gene [2]. $Na_{v1.5}$ is a large transmembrane voltage-gated channel made up of four repeated domains, with each domain comprising six subunits (S1–S6). S4 is positively charged and acts as the voltage sensor [2].

Movement of S4 can induce a conformational change in S5 and S6 that allows access to the pore-forming region through which Na^+ flows. Hence the configuration state of the channel, i.e. open or closed, is voltage-dependent.

The inactivation of $Na_{v1.5}$ occurs in both a voltage and time-dependent manner. Inactivation is the rapid block of Na^+ movement through the channel and must occur to allow the cell to repolarise. Once inactivated, channels must return to their closed state before their reactivation is possible. This change of state (from inactivated to closed) is achieved by repolarisation of the membrane [5]. The two-step process for reactivation is responsible for the action potential refractory period. The absolute refractory period is the time during which another action potential cannot be generated because the majority of Na^+ channels are in the inactive state and thus cannot be re-opened [5]. The relative refractory period occurs as the cells repolarise more so that a larger portion of Na^+ channels have recovered from inactivation and have returned to their closed state. Under these conditions a stronger than normal stimulus may depolarise the cell sufficiently to trigger channel opening that may lead to the production of a premature depolarisation [5].

5.2.1.1 Clinical Implications

Genetic mutations in the *SCN5A* gene may cause functional changes in the $Na_{v1.5}$ channel, as seen in type 3 Long QT and Brugada syndromes [6]. In both cases, the disruption of Na^+ channel inactivation leads to prolongation of the action potential, increasing the risk of arrhythmogenesis [6]. This is discussed further in ▶ Chap. 8.

Unlike many voltage-gated sodium channels in the human body, $Na_{v1.5}$ is not inhibited by low concentrations of tetrodotoxin; it is considered 'insensitive' to this compound [7]. Tetrodotoxin poisoning, occassionally a result of eating Japanese Fugu fish, is associated with respiratory arrest but few cardiac implications [7].

5.2.2 ɸ 1 – Early Repolarisation

Manifesting as the distinctive notch in ventricular APs, phase 1 results from the action of two transient outward currents, I_{to1} and I_{to2}, that summate to cause initial repolarisation of the cell [8]. This early repolarisation helps to prevent the cell depolarising to E_{Na}, and assists in the timely activation of channels involved in ɸ 2. Changes to the currents involved in this phase affects AP duration. The two transient outward currents are:

I_{to1} The efflux of potassium through voltage-gated K^+ channels generates a repolarising outward current. The main channels involved in humans are isoforms of the rapidly inactivating A-type K^+ channels, $K_{v4.2}$ and $K_{v4.3}$, coded by the *KCND2* and *KCND3* genes respectively [9]. Due to these voltage-gated channels undergoing both rapid activation and inactivation, I_{to1} only participates in early repolarisation. Inhibition of these channels using 4-aminopyridine is associated with action potential prolongation [10].

I_{to2} An outward current generated by the influx of Cl^- in a Ca^{2+}-mediated process, repolarises the cell towards E_{Cl}. The precise identity of these channels is uncertain but they open in response to changes in intracellular Ca^{2+} concentration [9]. Activation and inactivation times are slower compared with I_{to1}.

5.2.3 ɸ 2 – Plateau

During the plateau phase, the membrane potential changes very slowly due to a fine balance between inward Ca^{2+} currents and outward K^+ currents. The Na^+/Ca^{2+} exchange ion transporter (NCX) can contribute to the inward current because of the exchange stoichiometry of $3Na^+$ and $1Ca^{2+}$. Under normal physiological circumstances, during the plateau NCX will expel one Ca^{2+} ion from the myocyte in exchange for the inward movement of three Na^+ ions producing a net +1 inward current [5].

5.2.3.1 Inward Current (I_{Ca})

Following depolarisation, the membrane voltage range enables Ca^{2+} channel activation and therefore Ca^{2+} influx: an absolute requirement of myocyte contraction [11]. The activation of L-type Ca^{2+} channels (LTCCs), encoded by the *CACNA1C* gene, is responsible for the inward Ca^{2+} flux that occurs early in the AP and continues during the plateau [12]. Ca^{2+} influx triggers a complex process known as *'calcium-induced calcium release'*, which will be discussed in the ▶ Chap. 6. The L-type Ca^{2+} channel, also known as $Ca_{v1.2}$, is composed of 5 subunits, with the $\alpha1$ component containing the voltage sensor responsible for detecting membrane potential and forming the conducting pore [9, 12]. The ε subunit is responsible for the Ca^{2+}-dependent inactivation process.

LTCC inactivation is dependent on both membrane potential and intracellular Ca^{2+} concentration. Mechanistically, the voltage-dependent inactivation involves cytoplasmic loops locking with the cytoplasmic end of S6 to block calcium flux. In contrast, Ca^{2+}-dependent inactivation occurs as the sub-sarcolemmal Ca^{2+} concentration increases, due to influx across the surface membrane and preliminary release from the sarcoplasmic reticulum (SR). This form of inactivation presents a negative feedback loop [12].

Clinical Implications

The LTCC is inhibited by both dihydropyridine Ca^{2+} channel antagonists, e.g. nifedipine, and also non-dihydropyridine antagonists such as verapamil [12]. Although this drug class will be discussed in greater detail in ▶ Chap. 15, clinically, the effect of LTCC inhibition can produce antiarrhythmic and antihypertensive effects depending on the location of the channels being acted upon [13]. Dihydropyridines modulate LTCCs in the vasculature and therefore have antihypertensive effects. Non-dihydropyridines (also referred to as phenylalkylamines) modulate LTCCs located in the heart itself and have been demonstrated to exert antiarrhythmic effects [14].

5.2.3.2 Outward Currents (I_{Kur} I_{Kr} I_{Ks})

Depolarisation induces the gradual activation of a series of K^+ currents known as delayed rectifiers [9]:
- I_{Kur} – ultra rapid (present only in the atria)
- I_{Kr} – rapid
- I_{Ks} – slow

During the plateau, there is a delay in the onset of the opening of these channels, with outward I_{Kr} and I_{Ks} activating progressively as LTCCs inactivate. The combination of a decline in inward current and an increase in outward current repolarises the cell, eventually bringing the plateau to an end and starting the more rapid phase of repolarisation [5, 9]. Following the plateau, as membrane potential becomes more negative, a large increase in membrane permeability to K^+ occurs from the opening of a different type of K^+ channel that behaves in an unusual way (see Sect. 5.2.4) [5].

The delayed rectifier K^+ channels consist of four subunits that form a tetrameric structure with a central pore. Each subunit contains six transmembrane segments, with the S4 segment containing the voltage sensor whereby a change in voltage results in S4 rotation. This pulls on the S4-S5 linkage, triggering a conformational change in the S5-S6 'glycine hinge' that causes the channel to open [15]. This encapsulates the concept of voltage-gated channels, whereby a channel involves a gate and a sensing device that controls the gate by voltage. Multiple genes encode the wide variety of K^+ channel isoforms [15]. For the delayed rectifiers [9, 16]:

- *KCNA5* encodes the channel $K_{v1.5}$ producing I_{Kur}
- *KCNH2* encodes the channel $K_{v11.1}$ producing I_{Kr}
- *KCNQ1* encodes the channel K_{vLQT} producing I_{Ks}

Clinical Implications

Class III antiarrhythmic drugs are the principal K^+ channel inhibitors, specifically I_{Kr}. An example is amiodarone, which acts to block the K^+ channel, decreasing the channel conductance to this ion and inhibiting the outward flux of K^+ [17]. As a result, phase 2 of the AP is prolonged, increasing the duration of the refractory period. This therpeutic approach is commonly used to terminate arrhythmic re-entrant circuits. Importantly however, prolongation of the plateau phase predisposes to early afterdepolarisations (EADs), an abnormal depolarisation of the myocyte during phase 2 or 3 of the AP [17]. This illustrates the potential arrhythmic effect of antiarrhythmics in certain situations.

5.2.4 φ 3 – Fast Repolarisation

I_{K1} is the outward current responsible for the final rapid repolarisation, fully terminating the action potential and stabilising the resting membrane potential [9]. Often referred to as an *'inward rectifying channel'*, it is generated from the co-assembly of the $K_{ir}2.1.x$ subfamily of proteins ($K_{ir}2.1$ [main ventricular isoform], 2.2, and 2.3) [9, 18].

During the plateau phase [2] of the ventricular AP, a combination of the time-dependent inactivation of LTCCs and increasing delayed rectifier current activation results in a repolarisation of membrane potential to approximately -10 mV [5, 11].

At this voltage inward rectifier channels start to open increasing the magnitude of I_{K1} and thus K^+ conductance across the membrane [18]. This results in the membrane potential becoming more negative, increasing the probability of K_{ir} channel opening. With K_{ir} channels capable of passing large amounts of outward current, their activation quickly repolarises the membrane, hence the name 'fast repolarisation' [18]. Moreover, once the resting membrane potential has been reached, the channel produces an outward current that 'pulls' the membrane potential towards the K^+ equilibrium potential and establishes a stable potential [18].

It is of note that the channels that produce I_{K1} do not behave in the same way as many other K$^+$ channels and have a different structure. These channels comprise four subunits that form a tetramer with a central pore similar to K$_v$ channels [9]. Importantly however, K$_{ir}$2.x subunits differ in that each has two transmembrane segments, M1 and M2, instead of the six seen in K$_v$ subunits. With no voltage-gated S4 (typical of standard K$^+$ channels), these channels use a different mechanism to confer voltage sensitivity [9]. I_{KACh} and I_{K-ATP} are two other examples of inward rectifier currents with a similar channel structure to I_{K1}.

With inward rectifier channels lacking an S4 segment, K$^+$ movement through the channel is hindered (and facilitated) by the binding (and unbinding) of so-called 'blocking molecules' such as magnesium and voltage-sensitive (-10 to $20/30$ mV) polyamines [19]. Blocking molecules enter the channel pore to obstruct the efflux of K$^+$. At negative voltages less than -10 mV, the blocking molecules leave the channel pore, allowing the flow of potassium. For this reason, I_{K1} does not pass current during early repolarisation/phases 1 and 2, as the voltage is more positive than the -10 mV required for the displacement of blocking molecules [19].

5.2.5 ϕ 4 – Resting Membrane Potential

Phase 4, also referred to as 'resting membrane potential (RMP)', results primarily from I_{K1} and I_{KACh} currents, which together increase the membrane K$^+$ conductance. As previously discussed, these two currents conduct at negative E_m and mediate an efflux of K$^+$ that maintains polarisation of the cell in an attempt to reach the K$^+$ equilibrium potential, E_K. This high K$^+$ permeability is partially opposed by fluxes of other ions (e.g. some Na$^+$ and Ca^{2+} influx) across the membrane.

The consequence of this is that the RMP in cardiac myocytes does not reach E_K, and is slightly more positive [20]. Theoretically, if channels in the membrane were only permeable to K$^+$, the membrane potential would be equal E_K [5]. Importantly however, the foremost determinant of stabilising RMP is the passive flux of K$^+$ mediated by the high conductance I_{K1} [5].

Also of note is the activity of ATPase transporters, the most notable of which is the sodium-potassium pump (Na$^+$/K$^+$-ATPase). These consume ATP to actively move three Na$^+$ out of the cell against their concentration gradient whilst simultaneously moving two K$^+$ in. This process is essential for the maintenance of concentration gradients [5].

5.2.5.1 Other Currents

Late I_{Na} Carried by a subset of Na$^+$ channels that do not inactivate rapidly and produce a current that persists late into the AP. While this inward current is small in magnitude compared with the main Na$^+$ current, it can prolong the AP duration by opposing the outward current generated by delayed rectifiers. Its pathological effects are seen in heart failure, where its current is increased – a phenomenon that may derive from the hyperphosphorylation of Na$^+$ channels, increasing their open probability [21]. In these circumstances, the enhanced late I_{Na} is associated with the development of arrhythmias in two main ways [21]:

1. By prolonging the action potential. If AP duration increases, some LTCCs are able to recover from inactivation and will reopen because the correct voltage range is maintained. This results in Ca^{2+} influx into the cell, producing a substantial inward current that promotes EAD formation.

2. By increasing the intracellular Na^+ concentration. The enhanced late Na^+ current can cause delayed afterdepolarisations (DADs), because increased cytosolic Na^+ load leads to NCX reverse mode and Ca^{2+} influx.

I_{KACh} Acetylcholine (ACh) released by the parasympathetic vagus nerve binds to M2 muscarinic cholinergic receptors, triggering a G-protein cascade that activates $I_{K,ACh}$ channels, causing the efflux of K^+ ions and hyperpolarisation of the membrane [22]. This results in the threshold potential being reached more slowly, decreasing the AP firing rate and therefore heart rate [22].

$I_{K,ATP}$ Sarcolemmal ATP-sensitive K^+ (K_{ATP}) channels aid in protecting the heart against ischaemia via metabolo-electrical coupling [23]. Closed under normal circumstances, inhibition of cellular metabolism (such as during ischaemia) decreases cellular ATP concentration. This increases the open probability of K_{ATP} channels, augmenting K^+ efflux [23]. The AP duration is therefore shortened, decreasing contraction and preserving the cell's energy status. However, the decreased refractory period also provides a substrate for re-entry, promoting arrhythmogenesis [24].

$I_{K,ATP}$ activation has also been shown to limit Ca^{2+} entry during ischaemia by maintaining a negative and stable resting membrane potential, decreasing NCX-mediated Ca^{2+} influx and hence preventing Ca^{2+} overload [23, 24].

5.2.5.2 The I_f (Funny Current) in Cardiac Pacemaking

Cardiac pacemaking occurs from spontaneous diastolic depolarisations (DDs) that arise in the SAN. This relies on complex interactions between a nexus of interdependent currents, the precise mechanism of which remains unclear. That said, a vital current associated with cardiac pacemaking is I_f (the funny current), which is localised to pacemaking cells and is distinguished by three factors [25, 26]:

1. I_f occurs when cells are hyperpolarised.
2. I_f is a non-specific inward cationic current carried by K^+ and Na^+.
3. I_f flows through channels activated by both hyperpolarisation and cyclic AMP (cAMP). They are known as hyperpolarisation-activated, cyclic nucleotide-gated (HCN) channels.

In this way, variation in the concentration of cAMP facilitates modulation of cardiac pacemaking, and thus heart rate [27]. HCN is encoded by four genes (HCN1-4) with contrasting cAMP sensitivity and channel kinetics. HCN4 is the predominant SAN isotype, with high cAMP sensitivity and slow gating [28]. Despite having considerable structural commonality with voltage-gated potassium (K_v) channels, HCN4 displays higher sodium conductance at physiological voltages, facilitating a net inward current critical in DD [29]. Due to its activation by hyperpolarisation, this current constitutes the focal point of the so-called 'DiFrancesco' theory of cardiac pacemaking, named after its key proponent, Dario DiFrancesco [30].

5.3 What We Don't Know: The Mechanism of Cardiac Pacemaking

Two primary hypotheses exist for the spontaneous depolarisation observed in pacemaking cells: Dario DiFrancesco's I_f channel and Edward Lakatta's Ca^{2+} clock model. The first school of thought centres around the I_f channel, which is activated on hyperpolarisation and provides a pathway for the non-specific influx of cations (mainly Na$^+$ but also K$^+$) that depolarises the cell bringing the membrane potential closer to the activation potentials of T-and-L-type Ca^{2+} channels [30]. Evidence for this theory is provided by the contribution of I_f to spontaneous AP firing [30].

An alternative theory has been proposed by Lakatta et al., who suggest that spontaneous AP firing results from the summated interactions of two clocks: membrane (M) and Ca^{2+} [31]. The M-clock contains L-and-T-type Ca^{2+} currents, while the Ca^{2+} clock refers to SR ryanodine receptors (discussed in later chapters). Evidence for this theory is provided in the form of membrane potential investigations by Bogdanov et al., who demonstrate the critical period of exponential voltage rise during DD results from the synergistic interactions of membrane Ca^{2+} currents and sub-sarcolemmal Ca^{2+} release by ryanodine receptor (RyR) clusters [32]. The importance of this clock interplay is seemingly reinforced by studies implementing ryanodine-mediated block of RyR, causing a subsequent negative chronotropic effect in SA nodal cells [33, 34].

Overall however, there is a lack of definitive experiments that implicate one or other mechanism. Pacemaking may be, in fact, an amalgamation of both mechanisms, with I_f contributing to the membrane clock that triggers the Ca^{2+} clock and thereby spontaneous activity.

Take-Home Message

- The differences in action potential profile between ventricular, atrial and SA nodal cardiac myocytes arise from variance in the magnitude of specific ionic currents.
- This results from differences in the degree of expression of relevant ion channels.
- Changes in the above can have profound arrhythmogenic consequences.

Phase		Current	Direction	Channel
0	Upstroke	I_{Na}	Inward	Na$_{v1.5}$
1	Early repolarisation	I_{to1}	Outward	K$_{v4.3}$, K$_{v4.2}$
		I_{to2}	Outward	
2	Plateau	I_{Ca}	Inward	Ca$_{v1.2}$
		I_{Kur}	Outward	K$_{v1.5}$
		I_{Kr}	Outward	K$_{v11.1}$
		I_{Ks}	Outward	K$_{vLQT}$
3	Fast repolarisation	I_{K1}	Outward	K$_{ir}$2.1/2.2/2.3

References

1. Burkhard S, van Eif V, Garric L, Christoffels VM, Bakkers J (2017) On the evolution of the cardiac pacemaker. J Cardiovasc Dev Dis 4(2):4
2. Grant AO (2009) Cardiac ion channels. Circ Arrhythmia Electrophysiol 2(2):185–194

3. Klabunde RE (2012) Cardiovascular physiology concepts, 2nd edn. Lippincott Williams & Wilkins, Philadelphia
4. Hodgkin AL, Huxley AF (1952) A quantitative description of ion currents and its applications to conduction and excitation in nerve membranes. J Physiol 117:500
5. MacLeod K (2014) An essential introduction to cardiac electrophysiology. Imperial College Press, London
6. Wang Q, Shen J, Splawski I, Atkinson D, Li Z, Robinson JL et al (1995) SCN5A mutations associated with an inherited cardiac arrhythmia, long QT syndrome. Cell 80:805
7. Gellens ME, George AL, Chen LQ, Chahine M, Horn R, Barchi RL et al (1992) Primary structure and functional expression of the human cardiac tetrodotoxin-insensitive voltage-dependent sodium channel. Proc Natl Acad Sci 89:554
8. Kenyon JL, Gibbons WR (1979) Influence of chloride, potassium, and tetraethylammonium on the early outward current of sheep cardiac Purkinje fibers. J Gen Physiol 73(2):117–138
9. Bartos DC, Grandi E, Ripplinger CM (2015) Ion channels in the heart. Compr Physiol 5(3):1423–1464
10. Zygmunt AC (1992) Properties of the calcium-activated chloride current in heart. J Gen Physiol 99:391
11. Marks AR (2003 Mar) Calcium and the heart: a question of life and death. J Clin Invest 111(5):597–600
12. Bers DM, Perez-Reyes E (1999) Ca channels in cardiac myocytes: structure and function in Ca influx and intracellular Ca release. Cardiovasc Res 42:339
13. Stone PH, Antman EM, Muller JE, Braunwald E (1980 Dec) Calcium channel blocking agents in the treatment of cardiovascular disorders. Part II: hemodynamic effects and clinical applications. Ann Intern Med 93(6):886–904
14. Szentandrássy N, Nagy D, Hegyi B, Magyar J, Bányász T, Nánási PP (2015) Class IV antiarrhythmic agents: new compounds using an old strategy. Curr Pharm Des 21(8):977–1010
15. Snyders DJ (1999) Structure and function of cardiac potassium channels. Cardiovasc Res 42:377
16. Schmitt N, Grunnet M, Olesen S-P (2014) Cardiac potassium channel subtypes: new roles in repolarization and arrhythmia. Physiol Rev 94:609
17. Roden DM (2016) Pharmacogenetics of potassium channel blockers. Card Electrophysiol Clin 8:385
18. Dhamoon AS, Jalife J (2005) The inward rectifier current (IK1) controls cardiac excitability and is involved in arrhythmogenesis. Heart Rhythm 2:316
19. Liu T-A, Chang H-K, Shieh R-C (2012) Revisiting inward rectification: K ions permeate through Kir2.1 channels during high-affinity block by spermidine. J Gen Physiol 139:245
20. Santana LF, Cheng EP, Lederer WJ (2010) How does the shape of the cardiac action potential control calcium signaling and contraction in the heart? J Mol Cell Cardiol 49:901
21. Antzelevitch C, Nesterenko V, Shryock JC, Rajamani S, Song Y, Belardinelli L (2014) The role of late I Na in development of cardiac arrhythmias. Handb Exp Pharmacol 221:137–168
22. Mesirca P, Marger L, Toyoda F, Rizzetto R, Audoubert M, Dubel S et al (2013) The G-protein–gated K+ channel, I_{KACh}, is required for regulation of pacemaker activity and recovery of resting heart rate after sympathetic stimulation. J Gen Physiol 142(2):113–126
23. Tinker A, Aziz Q, Thomas A (2014) The role of ATP-sensitive potassium channels in cellular function and protection in the cardiovascular system. Br J Pharmacol 171:12–23
24. Flagg TP, Nichols CG (2011) Cardiac K_{ATP}. Circ Arrhythmia Electrophysiol 4(6):796–798
25. DiFrancesco D, Borer JS (2007) The funny current: cellular basis for the control of heart rate. Drugs 67(Suppl 2):15–24
26. DiFrancesco D, Tortora P (1991) Direct activation of cardiac pacemaker channels by intracellular cyclic AMP. Nature 351(6322):145–147
27. DiFrancesco D, Mangoni M (1994) Modulation of single hyperpolarization-activated channels (i(f)) by cAMP in the rabbit sino-atrial node. J Physiol 474(3):473–482
28. Moosmang S, Stieber J, Zong X, Biel M, Hofmann F, Ludwig A (2001 Mar 15) Cellular expression and functional characterization of four hyperpolarization-activated pacemaker channels in cardiac and neuronal tissues. Eur J Biochem 268(6):1646–1652
29. Robinson RB, Siegelbaum SA (2003) Hyperpolarization-activated cation currents: from molecules to physiological function. Annu Rev Physiol 65(1):453–480
30. Difrancesco D (2010) The role of the funny current in pacemaker activity. Circ Res 106:434

31. Maltsev VA, Lakatta EG (2008) Dynamic interactions of an intracellular Ca2+ clock and membrane ion channel clock underlie robust initiation and regulation of cardiac pacemaker function. Cardiovasc Res 77:274–284
32. Bogdanov KY, Vinogradova TM, Lakatta EG (2001) Sinoatrial nodal cell ryanodine receptor and Na(+)-Ca(2+) exchanger: molecular partners in pacemaker regulation. Circ Res 88(12):1254–1258
33. Rubenstein DS, Lipsius SL (1989) Mechanisms of automaticity in subsidiary pacemakers from cat right atrium. Circ Res 64(4):648–657
34. Satoh H (1997) Electrophysiological actions of ryanodine on single rabbit sinoatrial nodal cells. Gen Pharmacol 28(1):31–38

Cardiac Excitation-Contraction Coupling

Fotios G. Pitoulis and Cesare M. Terracciano

© Springer Nature Switzerland AG 2019
C. Terracciano, S. Guymer (eds.), *Heart of the Matter*, Learning Materials in Biosciences,
https://doi.org/10.1007/978-3-030-24219-0_6

What You Will Learn in This Chapter

This chapter will provide you with an understanding of the regulation of Ca^{2+} in the myocardium, its physiological implication as well as its role in orchestrating myocardial contraction. The chapter explores the processes of excitation-contraction coupling (ECC) and calcium-induced calcium release (CICR) whilst appreciating the relevance of ECC in pathology and in engineering heart tissue.

Learning Objectives

- Understand the molecular mechanism that underlie calcium-induced calcium release
- Appreciate and assess the different theories for the calcium-induced calcium release termination process
- Be able to discuss the issues with calcium handling in induced pluripotent stem cells

6

6.1 Introduction to Excitation-Contraction Coupling

Of the array of ions involved in the workings of the heart, calcium (Ca^{2+}) is perhaps the most important [1]. During the cardiac action potential, Ca^{2+} entry through the sarcolemmal Ca^{2+} channels stimulates Ca^{2+} release from the sarcoplasmic reticulum (SR), causing a rise in cytosolic Ca^{2+} and the subsequent activation of troponin on myofilaments (see ▶ Chap. 10), resulting in the development of force to eject blood out of the ventricles [2]. The process that links myocyte electrical excitation to contraction is known as excitation-contraction coupling (ECC). Appreciating ECC is crucial as it forms the basis of physiology, is dysregulated in almost all pathology and acts as a marker of the robustness of novel experimental cardiac models such as stem cell-derived cardiomyocytes [1, 3–5].

In each heartbeat, the cytoplasmic Ca^{2+} concentration of a healthy cardiomyocyte (CM) oscillates from $\cong 100$ nM to 1 μM [6]. Precise Ca^{2+} regulation is a matter of life and death, and improper cytoplasmic Ca^{2+} rise and/or removal can lead to defective systole and diastole, respectively (known as systolic and diastolic dysfunction).

6.2 Ca²⁺ Influx

During the ventricular action potential, influx of Ca^{2+} from the extracellular to subsarcolemmal space generates a Ca^{2+} current, known as I_{Ca}, which triggers Ca^{2+} release from the SR, mediated by SR-release channels known as ryanodine receptors (RyRs). I_{Ca} occurs in two main ways [5]:

- Voltage-sensitive sarcolemmal Ca^{2+} channels (LTCCs)
- Na^+-Ca^{2+} exchanger (NCX)

6.2.1 L-Type Ca²⁺ Channels (LTCCs)

LTCCs are activated by the initial membrane potential (V_m) depolarisation cause by the opening of voltage-gated Na^+ channels [1]. Following LTCC opening, deactivation occurs by both time-dependent, V_m-dependent, and cytosolic calcium ($[Ca^{2+}]_i$)-dependent mechanisms [5]. The V_m-dependent inactivation of the LTCCs can be demonstrated by administering depolarisation pulses and measuring LTCC inactivation kinetics [8].

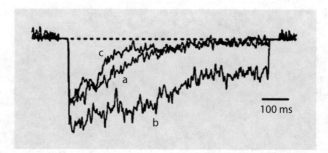

◘ Fig. 6.1 Ensemble currents demonstrating Ca^{2+}-dependent LTCC inactivation in planar lipid bilayer experiments; a, b, and c show currents with 10 µM-, 20 nM-, and 15 µM-[Ca^{2+}] respectively. Increasing concentrations of [Ca^{2+}] accelerate the rate of LTCC inactivation. (Image from [7])

Similarly, increasing [Ca^{2+}]$_i$ accelerates the inactivation rate of LTCCs, suggesting the precense of a negative-feedback system that prevents excess Ca^{2+} influx [7]. This is observed when Ca^{2+} is replaced with Ba^{2+} and the LTCC inactivation rate decelerates as Ca^{2+}-dependent inactivation is minimised [5].

Enzymatic and non-enzymatic mechanisms have been proposed to explain Ca^{2+}-dependent inactivation. In the former, dephosphorylation of the LTCC by Ca^{2+}-activated phosphatases deactivates the channel [7]. In the latter, a Ca^{2+}-calmodulin complex on the -COOH terminal of the α1 subunit of the LTCC binds Ca^{2+} when local [Ca^{2+}]$_i$ increases, altering the channel's conformation and thus inactivating it [5, 7] (◘ Fig. 6.1).

6.2.2 Na$^+$-Ca^{2+} Exchanger (NCX)

The second contributor of Ca^{2+} influx during the action potential is the Na$^+$-Ca^{2+} exchanger (NCX): a counter-transport system that operates by exchanging 3 Na$^+$ for 1 Ca^{2+}. This net movement of positive charge in the direction of Na$^+$ makes NCX electrogenic (i.e. it generates current). Typically, NCX moves Na$^+$ ions in whilst Ca^{2+} is effluxed out of the cell – known as the 'forward mode' and producing an inward current. NCX can also function in 'reverse mode', loading the cell with Ca^{2+} whilst Na$^+$ ions are effluxed out of the cell (outward current). This can be summarised mathematically in a few equations:

$$E_{rev} = E_{NCX} = 3E_{Na} - 2E_{Ca} \tag{6.1}$$

$$3(E_{Na} - V_m) > 2(E_{Ca} - V_m) \tag{6.2}$$

$$V_m < E_{NCX} \tag{6.3}$$

Let's work through those:

1. E_{rev} or E_{NCX} is the reversal potential of NCX – that is, the V_m at which NCX will switch from 'forward' to 'reverse' mode. As the equation shows, this depends on the individual equilibrium potentials of Na$^+$ and Ca^{2+}. This is exactly the same as the reversal potential of an ion channel, meaning that the current produced by NCX (I_{NCX}) when the membrane potential is equal to the reversal potential ($V_m = E_{NCX}$) is zero – there is no net movement of charge through the sarcolemma.

6

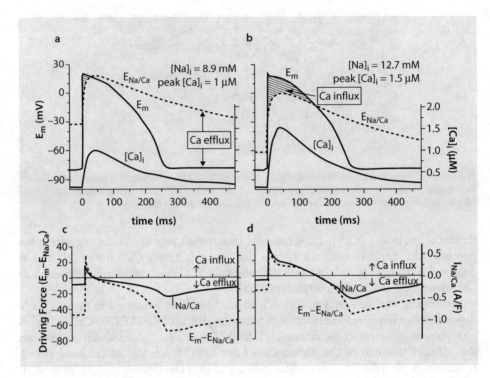

☐ **Fig. 6.2** Mode of operation of NCX: (**a, b**) V_m shown as E_m on graph, intracellular Ca^{2+} ([Ca^{2+}]$_i$) and E_{NCX}, with **a** low intracellular Na^+ ([Na^+]$_i$) and **b** high [Na^+]$_i$. Notice that with higher [Na^+]$_i$, $V_m > E_{NCX}$ for a greater period of time, meaning NCX functions in reverse mode for longer, promoting higher [Ca^{2+}]$_i$. **c** Driving force corresponds to the thermodynamic drive, which determines the mode of operation of NCX according to Eq. 6.2. When $V_m - E_{NCX} > 0$ there is Ca^{2+} influx due to the transporter operating in reverse mode. **d** Current generated by NCX (I_{NCX}). Notice that $I_{NCX} > 0$ (i.e. outward) when $V_m > E_{NCX}$ and $I_{NCX} < 0$ (i.e. inward) when $V_m < E_{NCX}$. Also, notice that when $E_{NCX} = V_m$, $I_{NCX} = 0$. ($I_{NCX} = I_{NA/Ca}$)(Image from [5])

2. This equation shows the thermodynamic basis for the transport that governs NCX, suggesting that when the energy for the inward movement of three Na^+ ions exceeds the energy for the inward movement of one Ca^{2+} ion, Na^+ influx and Ca^{2+} efflux are favoured (i.e. '*forward mode*'). Conversely, if $3(E_{Na} - V_m) < 2(E_{Ca} - V_m)$, then the reverse is thermodynamically favoured, and Ca^{2+} influx occurs [5].

3. This is a rearranged form of Eq. 6.2, demonstrating that when V_m is more negative than E_{NCX}, the exchanger functions in forward mode, and vice versa (☐ Fig. 6.2). That is:
   - $V_m < E_{NCX}$ – NCX operates in forward mode
   - $V_m > E_{NCX}$ – NCX operates in reverse mode

Ultimately, whether NCX promotes Ca^{2+} influx or efflux depends on its mode of operation, determined by (a) V_m, (b) E_{Na}, and (c) E_{Ca}. Thus, although intuitively it is sensible (and typically correct) to proclaim that when the subsarcolemmal Ca^{2+} is high NCX will favour Ca^{2+} extrusion, the mode of operation is not merely a function of Ca^{2+} (and by extension its equilibrium potential), but also E_{Na} and V_m. Evidently, the ability of NCX to operate bidirectionally makes it a pivotal player in Ca^{2+} homeostasis.

6.3 Ca²⁺ Efflux

The dissociation of Ca^{2+} from troponin on myofilaments allows relaxation to take place. For this to happen, Ca^{2+} must be removed from the cytoplasm. This occurs via four mechanisms:
1. Sarco/Endoplasmic Reticulum Ca^{2+}-ATPase (SERCA)
2. NCX (forward mode)
3. Sarcolemmal Ca^{2+}-ATPase
4. Mitochondrial Ca^{2+} transporters (into mitochondria)

6.3.1 SERCA Protein

Sarco/endoplasmic reticulum Ca^{2+}-ATPase (SERCA) is a protein pump concentrated on the longitudinal component of the SR, transporting Ca^{2+} from the cytoplasm to the SR lumen [9]. It has three different isoforms (SERCA1, 2 and 3), with SERCA2a expressed abundantly in the heart [10]. The transport reaction involves multiple steps, beginning with the binding of 2 Ca^{2+} ions and 1 ATP molecule on the pump's cytoplasmic side, in addition to phosphorylation. This triggers conformational alterations that facilitate the release of Ca^{2+} into the SR lumen, and H^+ into the cytoplasm [11]. Perhaps counterintuitively then, relaxation, and not merely contraction, is energy dependent [12].

In general, the ATP concentration required to saturate SERCA is 1000× fold lower than the cytoplasmic ATP of a healthy CM at any given time, meaning that except in the energy-starved heart (e.g. failing, dysrhythmic), lack of ATP is not the rate-limiting factor for Ca^{2+} removal [12]. However, ATP can also allosterically modulate SERCA activity via a lower affinity binding site on the pump, such that in ischaemia it is the lack of this allosteric effect, rather than of ATP available for hydrolysis, that may disrupt relaxation kinetics [5, 12].

The main SERCA2a activity regulator is a homopentameric protein known as phospholamban (PLN) [5, 13]. When PLN is dephosphorylated, it tonically inhibits SERCA by increasing its $K_m(Ca^{2+})$, meaning more Ca^{2+} is required to attain the same Ca^{2+} transport rate [14]. When phosphorylated (e.g. in response to adrenergic stimulation), this tonic inhibition is lifted, enhancing SERCA affinity for Ca^{2+} by decreasing its $K_m(Ca^{2+})$, thus accelerating Ca^{2+} SR sequestration and relaxation.

PLN phosphorylation is at least in part accountable for the positive lusitropic effects observed in the presence of adrenergic stimulation. Phosphorylation of PLN has been demonstrated in three sites including (a) serine-16, (b) threonine-17 and (c) serine-10 by cAMP-dependent protein kinase A, Ca^{2+}/Calmodulin protein kinase II and Ca^{2+}-activated protein kinase C, respectively [14]. Dephosphorylation of PLN by SR-associated phosphatases restores PLN's tonic inhibition [14].

Sarcolipin (SLN) (a PLN homologue) is another SERCA regulator, albeit less well understood. It has been suggested that when SLN is co-expressed with PLN, SERCA2a Ca^{2+} affinity decreases more than with PLN alone (◘ Fig. 6.3) [15, 16]. This may be due to the increased concentration of active (inhibitory) PLN monomers in the presence of SLN [17]. In particular, PLN is present either in homopentameric or monomeric form, the latter of which exhibits increased inhibitory activity [15]. When SLN is co-expressed, it holds PLN in a monomeric form (preventing it from polymerizing into pentamers),

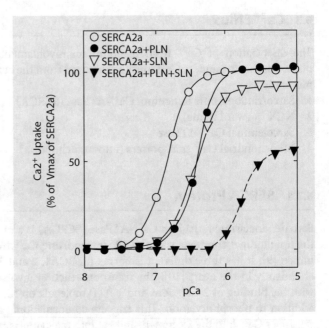

Fig. 6.3 Rate of Ca²⁺ uptake in HEK-293 cells expressing SERCA2a alone, SERCA2a and PLN, SERCA2a and SLN, SERCA2a and PLN and SLN. Note that the Ca²⁺ uptake rate is considerably decreased in the latter. (Image from [16])

6

enhancing its inhibitory activity and further decreasing SERCA Ca²⁺ uptake [15]. However, SLN has also been proposed to inhibit SERCA by a mechanism independent of PLN, as seen with SLN overexpression in PLN(⁻/⁻) (null mutants), which display impaired contractility, altered Ca²⁺ handling and relaxation [17, 18].

6.3.2 **NCX**

We previously considered the different modes of operation of NCX, demonstrating that when $V_m < E_{NCX}$, NCX extrudes Ca²⁺. In general, SERCA and NCX are the two main transporters responsible for relaxation, and at any one time are in the state of constant competition for Ca²⁺. This has important physiological implications and can be demonstrated by considering Ca²⁺ efflux in different species. For example, after a period of rest (i.e. no electrical stimulation), rabbit cardiac preparations show a decline in the amplitude of the first post-rest contraction, termed *rest decay* [19, 20] (Fig. 6.4).

In contrast, rat CMs have an increased contraction amplitude after a period of rest – known as post-rest potentiation [19]. Such phenomena are explained by the dynamic relationship of Ca²⁺ extruders. In both rabbits and rats during rest, Ca²⁺ leaks from the SR due to the random openings of RyRs (see *Ca²⁺ sparks*, below). Ca²⁺ is then subjected to two opposing forces – SERCA and NCX. In rabbits, NCX moves Ca²⁺ out of the cell, progressively decreasing the SR Ca²⁺ content, which results in less Ca²⁺ available for myofilament activation and a diminished contraction post-rest [5, 19, 20]. In contrast, rats have high intracellular Na⁺ (resulting in $V_m > E_{NCX}$), slow NCX, and fast SERCA transport rates, the summation of which leads to increased SR Ca²⁺ content during rest and a potentiated contraction post-rest [19].

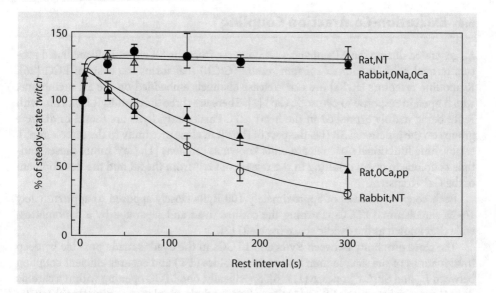

◻ Fig. 6.4   Rest decay and rest-potentiation in rabbits and rats, respectively. After a period of electrical rest for 100 s, the first contraction after rest is decreased to 75% of contraction during steady state in rabbits. In contrast, rats exhibit post-rest potentiation, increasing their contraction amplitude to approximately 140% after a period of electrical rest. Notice that thermodynamic inhibition of NCX in rabbits (0Na, 0Ca) converts the post-rest decay into post-rest potentiation. Similarly, applying a $0Ca^{2+}$ extracellular solution in rats favours Ca^{2+} extrusion by NCX and converts the typical post-rest potentiation into post-rest decay (NT = Normal Tyrode's solution). (Image from [19])

6.3.3   Other Extrusion Mechanisms

When caffeine is applied on cardiac preparations Ca^{2+} stored in the SR is released. If NCX and SERCA are blocked, the rate of cytoplasmic Ca^{2+} removal is significantly slowed but not completely abolished [21].

This is because in addition to NCX and SERCA, Ca^{2+} removal also occurs by sarcolemmal Ca^{2+}-ATPase and the mitochondrial Ca^{2+} uniporter (MCU). Sarcolemmal Ca^{2+}-ATPase utilises ATP to efflux Ca^{2+} out of the cell whilst MCUs facilitate the flux of Ca^{2+} in mitochondria down a large electrochemical gradient [22, 23]. With the exception of a few species (e.g. ferret), the contribution of these in removing Ca^{2+} on a beat-to-beat basis is marginal when compared to NCX or SERCA [24]. As such, they are known as *slow extruders*.

However, their role in maintaining CM health and function is anything but marginal. For instance, increased mitochondrial Ca^{2+} loading activates energy production (e.g. via ATP synthase) allowing CMs to cope with increased energy demands, yet prolonged periods of elevated intracellular Ca^{2+} can trigger mitochondrial dysfunction and acute CM death [25]. Ultimately, the rate of Ca^{2+} removal from the CM cytosol can be quantified to highlight the rate of removal by each transporter, such that the total rate of Ca^{2+} removal is equal to:

$$\frac{d[Ca]_t}{dt} = J_{SERCA} + J_{NCX} + J_{slow}$$

(6.4)

where J is the rate of removal of extrusion components, each governed by a nonlinear function dependent on Ca^{2+} concentration [5].

6.4 Excitation-Contraction Coupling

I_{Ca} generated during the AP enhances Ca^{2+} release from the SR via an RyR-mediated process termed *calcium-induced calcium release* (CICR) that forms the basis of ECC [26]. Ryanodine receptors (RyRs) are Ca^{2+} release channels embedded on the SR membrane, which open in response to cytosolic Ca^{2+} [1]. There are three different RyR isoforms, with RyR2 being mainly expressed in the heart [27]. Particularly, RyRs are found in discrete groups on the junctional SR (i.e. the part of the SR in close proximity to the sarcolemma), establishing functional Ca^{2+}-release units known as *couplons* [1]. Ca^{2+} influx causes multiple couplons to open, resulting in the release of Ca^{2+} from the SR and the development of the Ca^{2+}-transient.

Each couplon consists of approximately 100 RyRs closely apposed to approximately 10–25 sarcolemmal LTCCs, forming the cardiac dyad and separated by a nanometres-wide cleft known as the dyadic space (or cleft) [2].

The close proximity between RyRs and LTCCs in the dyad is made possible by deep invaginations of the sarcolemma known as t-tubules (TT) and ensures efficient coupling between I_{Ca} and SR-Ca^{2+} release [1, 2, 6]. Specifically, one LTCC opening within a cardiac dyad is sufficient to trigger SR-Ca^{2+} release from a whole couplon, meaning that the $\cong$10–25 LTCCs for $\cong$100 RyRs ensures a safety margin for triggering SR-Ca^{2+} release [26–28]. In addition to the role of cytosolic Ca^{2+} in triggering Ca^{2+} release via RyRs, SR luminal Ca^{2+} also plays a pivotal role in SR-Ca^{2+} release [29]. For instance, increased SR Ca^{2+} content can stimulate Ca^{2+} release, whilst RyR2 activity is diminished as luminal Ca^{2+} decreases [29].

6.4.1 Calcium Sparks

In 1993, Cheng et al. used fluorescent Ca^{2+} indicators and laser scanning confocal microscopy to describe the concept of Ca^{2+} sparks for the first time [26]. Ca^{2+} sparks are microscopic elevations of cytoplasmic Ca^{2+} reflecting the synchronous opening of a cluster of RyRs [1]. They occur by either of two mechanisms described below:

1. Stochastic openings of RyRs [26–30]. The open probability of a single or a small number of RyRs can randomly become non-zero, triggering nano-elevations of the resting $[Ca^{2+}]_i$ to about 170 nM [26]. This is important, as in pathological states associated with Ca^{2+} overload (supranormal SR Ca^{2+}-content), spontaneous SR Ca^{2+} release events can cause spark-induced spark-release, leading to high $[Ca^{2+}]_i$ ("macrosparks", $\cong$ 500 nM) and the successive development of dysrhythmogenic waves [26, 28].
2. Evoked by I_{Ca} that raises local subsarcolemmal Ca^{2+} and activates RyRs [26, 30]. During the cardiac action potential, I_{Ca} evokes multiple Ca^{2+} sparks by stochastically activating clusters of RyRs. The spatiotemporal summation of $\cong$10^4 individual Ca^{2+} sparks results in the production of the seeming spatially uniform Ca^{2+} transient [28].

6.4.2 Calcium-Induced Calcium Release

The idea that the Ca^{2+} transient consists of many individual 'atomic' subevents (i.e. Ca^{2+} sparks) revolutionised the ECC paradigm. Previously, many models of cardiac CICR assumed a common pool of cytosolic Ca^{2+}, consisting of homogenously distributed Ca^{2+} from I_{Ca} and SR Ca^{2+} release, with the former controlling the latter [31].

According to that model, when I_{Ca} stimulates SR Ca^{2+} release and the pool begins to fill with Ca^{2+}, a positive feedback loop is established in which Ca^{2+} released from the SR triggers more SR-Ca^{2+} release [31]. This makes SR-Ca^{2+} release an all-or-none response such that once the SR-Ca^{2+} release process commences, CICR is expected to evolve autonomously, irrespective of sarcolemmal Ca^{2+} influx [32].

However, experimental evidence does not support the notion that SR-Ca^{2+} release becomes autonomous. Instead, it is accepted that the magnitude of SR Ca^{2+}-release is a function of I_{Ca} [31, 33]. As I_{Ca} is primarily carried by the LTCCs, SR Ca^{2+} release is dependent on membrane potential [34]. If I_{Ca} is abruptly terminated by depolarisation above the LTCC reversal potential, SR Ca^{2+} release is also terminated [31, 35]. Therefore, SR Ca^{2+} release is graded – meaning it is a function of Ca^{2+} influx through LTCCs (I_{Ca}) (i.e. dependent on duration and magnitude of Ca^{2+} entry) [31, 36]. This is depicted in ◘ Fig. 6.5, which shows the typical characteristic bell-shaped voltage dependency of (a) I_{Ca} and (b) cell shortening (reflecting Ca^{2+} transient magnitude). Such a graded response is not in accordance with common pool models, which would be expected to cause an 'all-or-none' SR-Ca^{2+} release.

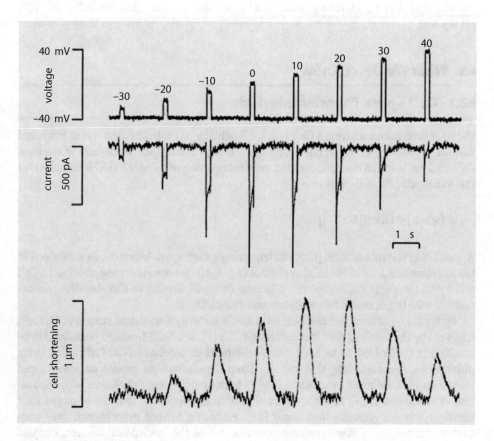

◘ **Fig. 6.5** Bell shaped LTCC current (I_{Ca}) in response to increasing membrane potential (voltage-clamp). Notice bell-shaped relationship of cell-shortening as well, reflecting bell-shaped Ca^{2+} transient amplitude (peaking at approximately 10 V)

To explain this gradation, Stern et al. proposed the local control theory of ECC, whereby Ca^{2+} sensed by RyRs is not the same as the average cytoplasmic $[Ca^{2+}]_i$ [33]. In particular, the opening of LTCCs causes a very high and rapid local rise of $[Ca^{2+}]$ within a cardiac dyad to >10 μM [33]. This activates RyRs within a couplon, causing SR-Ca^{2+} release to further elevate local $[Ca^{2+}]_i$ [33].

It is also proposed that the sensitivity of RyRs to Ca^{2+} is much less than the ambient cytosolic Ca^{2+}, preventing an 'all-or-none' regenerative calcium release [33]. Therefore, although CICR may be regenerative within an individual couplon (i.e. Ca^{2+} released by one RyR in a couplon triggering Ca^{2+} release by other RyRs in the same couplon – a positive feedback loop), Ca^{2+} released from one couplon does not spread in sufficiently high amounts to trigger Ca^{2+} release from neighbouring couplons [36].

These Ca^{2+} release events triggered by individual stochastic openings of LTCCs are in essence Ca^{2+} sparks, the spatial and temporal summation of which leads to the whole-cell Ca^{2+} transient [30]. Gradation of Ca^{2+} transient then occurs by the stochastic recruitment of more or less Ca^{2+} sparks according to the membrane potential (and by extension I_{Ca}) [34]. Ultimately, the distinction between common pool and local control models of ECC is highlighted by the fact that in the latter, elementary Ca^{2+}-sparks are recruited not by the mean $[Ca^{2+}]_i$ in the cell, but rather by the amount of Ca^{2+} flowing through the sarcolemmal LTCC, elevating local $[Ca^{2+}]_i$ in the cardiac cleft nanodomain [37] (◘ Fig. 6.6).

6.5   What We Don't Know

6.5.1 Ca^{2+} Spark Theories Interlude

But what determines whether a Ca^{2+} spark will actually be evoked? Santana et al. proposed that the probability of triggering a Ca^{2+} spark is dependent upon the square of the local Ca^{2+} concentration in the nanodomain, and that opening of a single LTCC is sufficient for this to happen [30, 38] – that is:

$$P(\text{spark}) = \left(\text{local}\left[Ca^{2+}\right]\right)^2$$

Appreciating the role of local $[Ca^{2+}]$ in triggering a Ca^{2+} spark is critical, as alterations in the microarchitecture of the ECC apparatus (e.g. in the geometric arrangement of LTCCs and RyRs) seen with pathology, can affect the $P(\text{spark})$, leading to Ca^{2+} handling abnormalities with implications for cardiac contractility [30].

By far the most accepted mechanism of CICR is the RyR-mediated release of SR Ca^{2+}, triggered by the influx of Ca^{2+} through the LTCCs [1]. Yet, NCX has been postulated to be involved in CICR [39]. In isolated CMs, LeBlanc et al. blocked LTCC Ca^{2+} influx using nisoldepine, demonstrating that voltage-clamp depolarisations caused an initial rapid inward current, followed by a rise in $[Ca^{2+}]_i$, both of which were abolished with application of tetrodotoxin (TTX, a Na^+ channel inhibitor), suggesting that the observed Ca^{2+} transient was Na^+ channel-dependent [39]. Following further experiments, they concluded that the initial depolarisation upstroke due to the inward Na^+ current, coupled with the increasingly positive V_m, promotes transient 'reverse mode' in NCX operation, leading to Ca^{2+} influx and providing the trigger for CICR [9, 39].

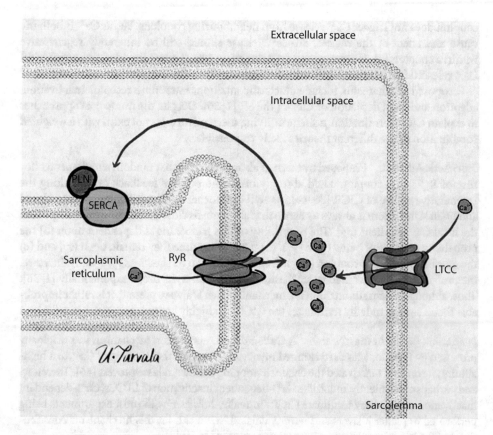

□ **Fig. 6.6** Close apposition of sarcolemmal LTCCs and RyRs on the junctional SR promote efficient coupling. During an AP, I_{Ca} increases $[Ca^{2+}]_i$ in the cardiac dyadic cleft, evoking Ca^{2+} sparks via RyR activation. Local $[Ca^{2+}]_i$ in the cardiac dyadic cleft is much higher than $[Ca^{2+}]$ in the bulk cytosolic space, allowing for RyR activation in couplons despite their low Ca^{2+} sensitivity

Another example comes from $NCX^{-/-}$ isolated ventricular myocytes. These cells display normal ECC, however in the presence of heavy Ca^{2+} buffering (minimising the effect of Ca^{2+} influx from LTCCs), reduced coupling efficiency is observed vs. wild-type CMs. This suggests there is an increased proportion of couplons failing to activate during the AP in the $NCX^{-/-}$ myocytes compared to wild type [40]. Accordingly, Goldhaber et al. proposed that NCX has a role in maintaining coupling during depolarisation by priming the dyadic space with a subthreshold amount of Ca^{2+}, meaning only a small amount of further Ca^{2+} from LTCCs is required to trigger CICR [40]. Others have remained sceptical of the role of NCX in ECC [41, 42], as (a) Na^+ channels may be excluded from the dyadic cleft, and (b) NCX as a transporter (and not an ion channel) is notably slower than LTCCs, meaning that when both co-exist, CICR is dominated by the latter [1, 43].

6.5.2 CICR Termination – Stopping the Domino Effect

We have seen that couplons are separated from each other and that according to the local control theory of ECC, RyR Ca^{2+} sensitivity is low enough so that Ca^{2+} released from one couplon does not trigger Ca^{2+} release from neighbouring couplons. Yet, as Ca^{2+} is both the cause and effect of the release, SR-Ca^{2+} release should still be inherently regenerative within a couplon [5]. However, with approximately 50% (i.e. not the whole amount) of SR Ca^{2+} released in each contraction, what terminates the release of Ca^{2+} from the SR?

Proposed mechanisms include stochastic attrition, ryanodine receptor inactivation, adaption and local depletion of Ca^{2+} in the SR [1, 36]. Despite the number of approaches to explain CICR termination, a single unifying mechanism does not exist, with a weighted combination of the different theories likely responsible.

Stochastic Attrition   Proposed by Stern et al. and suggests that random simultaneous closure of RyRs in a couplon could abruptly break the positive feedback cycle, halting the regenerative nature of CICR [36, 44]. As RyR are stochastically oscillating between closed and open states, there is always a chance that all channels close at the same time, degrading the local Ca^{2+} gradient [36]. The probability of this happening is dependent upon (a) the probability of the RyR being open (P_o), (b) the average time they remain open (τ_o) and (c) the number of RyR in a couplon (n) [36]. Generally, it can be shown as P_o, τ_o and n increase, the probability of simultaneous, stochastic closure of all RyRs drops exponentially [1, 36]. Thus, although stochastic attrition is a mechanism that is always present, it is rather improbable that it singlehandedly terminates the CICR of a highly active calcium synapse [36].

Ryanodine Receptor Inactivation   A Ca^{2+}-induced inactivation mechanism was originally proposed by Fabiato, who used skinned myocytes to suggest that binding of Ca^{2+} to a high-affinity site on RyR inactivated the channel, stopping the Ca^{2+} release process [45]. Therefore, and perhaps conveniently, much like Ca^{2+}-dependent inactivation of LTCCs, Ca^{2+}-dependent inactivation of RyR may terminate CICR. In reality however, with most experiments being performed in planar bilayers and never validated in intact cells, this mechanism's contribution to SR-Ca^{2+} release termination remains unclear [36, 45].

Adaption   RyRs relax to a lower P_o after activation. Strictly speaking, adaption is not RyR inactivation per se as the channels can be reactivated with exposure to higher $[Ca^{2+}]_i$ [1, 45].

Depletion of Local SR Luminal Ca^{2+}   Opening of RyRs could lead to depletion of SR luminal Ca^{2+} near the channel, resulting in either the flux of Ca^{2+} out of the SR to become zero (i.e. there is no more Ca^{2+} to be released), halting the positive feedback loop and ending CICR; or, causing the probability of RyR opening (in part determined by luminal Ca^{2+}) to be decreased such that no more Ca^{2+} is released [36]. As cytosolic Ca^{2+} is constantly being pumped into the SR, typically at a different location than the one releasing Ca^{2+}, the validity of this mechanism is dictated by (a) the rate of Ca^{2+} reuptake and (b) the rate of diffusion of Ca^{2+} from the 'uptake SR compartment' to the 'SR release compartment' [36].

6.6 Where We're Heading

6.6.1 ECC as a Measure of Cardiac Maturity

Stemcell-derived cardiomyocytes hold significant therapeutic promise for the treatment of diseased myocardium. These are stem cells that have been differentiated in to cardiomyocytes. A problem that has beset this field is the immaturity of such cells, in particular the absence of adult-like functional and morphological cardiomyocyte characteristics [46].

Components of the ECC seem particularly affected, with stem cell-derived cardiomyocytes having few or even no t-tubules, which can at least in part explain the abnormal Ca^{2+} dynamics and considerably weaker contractile force amplitudes than those developed by the adult myocardium (e.g. human ventricular strips twitch tension is $\cong 44$ mN/mm^2 vs. 0.08 mN/mm^2 in human pluripotent stem cell-derived cardiomyocytes – a 550-fold decrease) [46, 47].

In cell therapy, stem cell-derived cardiomyocytes are transplanted onto the injured myocardium in an effort to increase the contractile ability of the heart. Lack of robust ECC apparatus questions whether beneficial effects (if any) of such approaches reflect the addition of force-generating cardiomyocytes or merely the release of nurturing paracrine mediators from these cells [48].

Furthermore, transplantation of immature cells that beat asynchronously, have improper Ca^{2+} homeostasis, and/or are comparatively weaker to human cardiomyocytes may increase the dysrhythmogenic risk and by extension the safety of such approaches [4].

6.6.2 ECC in Pathology

Abnormal ECC underlies many pathological processes. In the failing heart, characterized by an inability to maintain a cardiac output sufficient to meet the metabolizing needs of the body [49, 50], disrupted geometrical arrangements of the ECC components may compromise the fidelity of ECC [51, 52]. For example, among the most well-known phenotypical changes seen in failing cardiomyocytes is the loss of t-tubules [53, 46]. Such morphological changes can disrupt the tight coupling between the LTCC and RyRs, and diminish the ability of a given I_{Ca} to trigger a Ca^{2+} spark, (and by extension a Ca^{2+} transient), hampering contractile performance [4, 52, 53]. Ingenious computational models have shown that for any spatial arrangement of LTCC and RyRs, there is an optimal amount of Ca^{2+} influx required to maximally activate RyRs [31]. If that becomes suboptimal (e.g. excessive Ca^{2+} influx which only minimally activates RyRs) in pathology then renormalizing this relationship between Ca^{2+} influx and RyR response with drugs or interventions may be of therapeutic benefit by enhancing contraction while simultaneously minimizing supranormal Ca^{2+} influx. This is important as excess Ca^{2+} influx raises the probabilities of Ca^{2+}-induced pathological states (e.g. Ca^{2+} overload-induced arrhythmias & activation of Ca^{2+} mediated pathological hypertrophy signalling pathways).

> **Take-Home Message**
>
> - During ventricular action potentials, Ca^{2+} influx from the extracellular to subsarcolemmal space generates a Ca^{2+} current, I_{Ca}, which initiates Ca^{2+} release from the sarcoplasmic reticulum via a process known as calcium-induced calcium release.
> - I_{Ca} is generated by two main mechanisms: voltage-sensitive sarcolemmal Ca^{2+} channels (LTCCs) and Na^+/Ca^{2+} exchangers (NCX).
> - Calcium-induced calcium release is mediated by ryanodine receptors interacting in complex microdomains with LTCCs.
> - The therapeutic usefulness of transplanting stem cell-derived cardiomyocyte onto injured myocardium is hindered in part by the deficient Ca^{2+} handling exhibited by these cells.

References

1. Bers DM (2002) Cardiac excitation-contraction coupling. Nature 415:198–205
2. Eisner DA, Caldwell JL, Kistamás K, Trafford AW (2017) Calcium and excitation-contraction coupling in the heart. Circ Res 121:181–195
3. Ibrahim M, Al Masri A, Navaratnarajah M, Siedlecka U, Soppa GK, Moshkov A et al (2010) Prolonged mechanical unloading affects cardiomyocyte excitation-contraction coupling, transverse-tubule structure, and the cell surface. FASEB J 24(9):3321–3329
4. Kane C, Couch L, Terracciano CMN (2015) Excitation–contraction coupling of human induced pluripotent stem cell-derived cardiomyocytes. Front Cell Dev Biol 3:59
5. Bers DM. Excitation-contraction coupling and cardiac contractile force. Springer Science; 1991
6. Marks AR (2003) Calcium and the heart: a question of life and death. J Clin Investig 111:597–600
7. Haack JA, Rosenberg RL (1994) Calcium-dependent inactivation of L-type calcium channels in planar lipid bilayers membrane preparation planar lipid bilayers. Biophys J 66(April):1051–1060
8. Zhang JF, Ellinor PT, Aldrich RW, Tsien RW (1994) Molecular determinants of voltage-dependent inactivation in calcium channels. Nature 372:97–100
9. Barry WH, Bridge JH (1993) Intracellular calcium homeostasis in cardiac myocytes. Circulation 87(6):1806–1815
10. Periasamy M, Kalyanasundaram A (2007) SERCA pump isoforms: their role in calcium transport and disease. Muscle Nerve 35:430
11. MacLennan DH, Green NM (2000) Pumping ions. Nature 405:633–634
12. Katz a M, Lorell BH (2000) Regulation of cardiac contraction and relaxation. Circulation 102(20 Suppl 4):IV69–IV74
13. Gustavsson M, Verardi R, Mullen DG, Mote KR, Traaseth NJ, Gopinath T et al (2013) Allosteric regulation of SERCA by phosphorylation-mediated conformational shift of phospholamban. Proc Natl Acad Sci 110(43):17338–17343
14. Frank KF, Bolck B, Erdmann E, Schwinger RHG (2003) Sarcoplasmic reticulum Ca2+ -ATPase modulates cardiac contraction and relaxation. Cardiovasc Res 57(April):20–27
15. MacLennan DH, Asahi M (2003) Tupling a R. The regulation of SERCA-type pumps by phospholamban and sarcolipin. Ann N Y Acad Sci 986(1):472–480
16. Asahi M, Kurzydlowski K, Tada M, MacLennan DH (2002) Sarcolipin inhibits polymerization of phospholamban to induce superinhibition of sarco(endo)plasmic reticulum Ca2+-ATPases (SERCAs). J Biol Chem 277(30):26725–26728
17. Periasamy M, Bhupathy P, Babu GJ (2008) Regulation of sarcoplasmic reticulum Ca2+ ATPase pump expression and its relevance to cardiac muscle physiology and pathology. Cardiovasc Res 77(2):265–273
18. Gramolini AO, Trivieri MG, Oudit GY, Kislinger T, Li W, Patel MM et al (2006) Cardiac-specific overexpression of sarcolipin in phospholamban null mice impairs myocyte function that is restored by phosphorylation. Proc Natl Acad Sci U S A 103(7):2446–2451
19. Bassani RA, Bers DM (1994) Na-ca exchange is required for rest-decay but not for rest-potentiation of twitches in rabbit and rat ventricular myocytes. J Mol Cell Cardiol 26:1335–1347
20. Bers DM (1991) Ca regulation in cardiac muscle. Med Sci Sport Exerc 23(10):1157–1162
21. Bassani RA, Bassani JW, Bers DM (1992) Mitochondrial and sarcolemmal Ca2+ transport reduce [Ca2+]i during caffeine contractures in rabbit cardiac myocytes. J Physiol 453(1):591–608
22. Choi HS, Eisner DA (1999) The role of sarcolemmal Ca^{2+}-ATPase in the regulation of resting calcium concentration in rat ventricular myocytes. J Physiol 515(1):109–118
23. Kirichok Y, Krapivinsky G, Clapham DE (2004) The mitochondrial calcium uniporter is a highly selective ion channel. Nature 427(6972):360–364
24. Bassani RA, Bassani JWM, Bers DM (1995) Relaxation in ferret ventricular myocytes: role of the sarcolemmal Ca ATPase. Pflugers Arch Eur J Physiol 430(4):573–578
25. Kwong JQ, Lu X, Correll RN, Schwanekamp JA, Vagnozzi RJ, Sargent MA et al (2015) The mitochondrial calcium uniporter selectively matches metabolic output to acute contractile stress in the heart. Cell Rep 12(1):15–22
26. Cheng H, Lederer W, Cannell M (1993) Calcium sparks: elementary events underlying excitation-contraction coupling in heart muscle. Science 262(5134):740–744
27. Lehnart SE, Wehrens XHT, Kushnir A, Marks AR (2004) Cardiac ryanodine receptor function and regulation in heart disease. Ann N Y Acad Sci 1015:144

28. Cheng H, Lederer WJ (2008) Calcium Sparks. Physiol Rev 88(4):1491–1545
29. Györke S, Terentyev D Modulation of ryanodine receptor by luminal calcium and accessory proteins in health and cardiac disease. Cardiovasc Res 2008, 77(2):245–255
30. Cannell MB, Soeller C (1998) Sparks of interest in cardiac excitation-contraction coupling. Trends Pharmacol Sci 19:16–20
31. Stern MD (1992) Theory of excitation-contraction coupling in cardiac muscle. Biophys J 63(2):497–517
32. Cannell MB, Kong CHT (2012) Local control in cardiac E-C coupling. J Mol Cell Cardiol 52:298
33. Stern MD, Song LS, Cheng H, Sham JS, Yang HT, Boheler KR et al (1999) Local control models of cardiac excitation-contraction coupling. A possible role for allosteric interactions between ryanodine receptors. J Gen Physiol 113(3):469–489
34. Hinch R, Greenstein JL, Tanskanen a J, Xu L, Winslow RL (2004) A simplified local control model of calcium-induced calcium release in cardiac ventricular myocytes. Biophys J 87(6):3723–3736
35. Barcenas-Ruiz L, Wier WG (1987) Voltage dependence of intracellular [ca 2+]i transients in guinea pig ventricular myocytes. Circ Res 61:148–154
36. Stern MD, Cheng H (2004) Putting out the fire: what terminates calcium-induced calcium release in cardiac muscle? Cell Calcium 35(6):591–601
37. Wier WG, Balke CW (1999) Ca2+ release mechanisms, Ca2+ sparks, and local control of excitation-contraction coupling in normal heart muscle. Circ Res 85(9):770–776
38. Santana LF, Cheng H, Gómez AM, Cannell MB, Lederer WJ, Scott JD et al (1996) Relation between the sarcolemmal Ca2+ current and Ca2+ sparks and local control theories for cardiac excitation-contraction coupling. Circ Res 78(1):166–171
39. Hume LN (1990) Sodium current-induced release of calcium from cardiac sarcoplasmic reticulum. Science 248(4953):372–376
40. Goldhaber JI, Philipson KD (2013) Cardiac sodium-calcium exchange and efficient excitation-contraction coupling: implications for heart disease. Adv Exp Med Biol 961:355–364
41. López-López JR, Shacklock PS, Balke CW, Wier WG (1995) Local calcium transients triggered by single L-type calcium channel currents in cardiac cells. Science 268(5213):1042–1045
42. Bers DM, Lederer WJ, Berlin JR (1990) Intracellular Ca transients in rat cardiac myocytes: role of Na-Ca exchange in excitation-contraction coupling. Am J Physiol Physiol 258(5):C944–C954
43. Sham JSK, Cleemann L, Morad M (1992) Gating of the cardiac Ca2+ release channel: the role of Na+ current and Na+ -Ca2+ exchange. Science 255:850–853
44. Sobie EA, Duly KW, Cruz JDS, Lederer WJ, Jafri MS (2002) Termination of cardiac Ca2+ sparks: an investigative mathematical model of calcium-induced calcium release. Biophys J 83(1):59–78
45. Sham JS, Song LS, Chen Y, Deng LH, Stern MD, Lakatta EG et al (1998) Termination of Ca2+ release by a local inactivation of ryanodine receptors in cardiac myocytes. Proc Natl Acad Sci U S A 95(25):15096–15101
46. Yang X, Pabon L, Murry CE (2014) Engineering adolescence: maturation of human pluripotent stem cell-derived cardiomyocytes. Circ Res 114:511–523
47. Mannhardt I, Breckwoldt K, Letuffe-Brenière D, Schaaf S, Schulz H, Neuber C et al (2016) Human engineered heart tissue: analysis of contractile force. Stem Cell Reports 7:29
48. Malliaras K, Marbán E (2011) Cardiac cell therapy: where weve been, where we are, and where we should be headed. Br Med Bull 98(1):161–185
49. Ponikowski P, Voors AA, Anker SD, Bueno H, Cleland JGF, Coats AJS et al (2016) 2016 ESC guidelines for the diagnosis and treatment of acute and chronic heart failure. Eur Heart J 37:2129–2200m
50. Zima AV, Bovo E, Mazurek SR, Rochira JA, Li W, Terentyev D (2014) Ca handling during excitation-contraction coupling in heart failure. Pflugers Archiv Eur J Physiol 466:1129–1137
51. Gomez AM, Valdivia HH, Cheng H, Lederer MR, Santana LF, Cannell MB et al (1997) Defective excitation-contraction coupling in experimental cardiac hypertrophy and heart failure. Science 276(5313): 800–806
52. Ibrahim M, Terracciano CM (2013) Reversibility of T-tubule remodelling in heart failure: mechanical load as a dynamic regulator of the T-tubules. Cardiovasc Res 98:225–232
53. Lyon AR, MacLeod KT, Zhang Y, Garcia E, Kanda GK, Lab MJ et al (2009) Loss of T-tubules and other changes to surface topography in ventricular myocytes from failing human and rat heart. Proc Natl Acad Sci U S A 106(16):6854–6859 ·

Conduction in Normal and Diseased Myocardium

Alec Saunders and Fu Siong Ng

© Springer Nature Switzerland AG 2019
C. Terracciano, S. Guymer (eds.), *Heart of the Matter*, Learning Materials in Biosciences,
https://doi.org/10.1007/978-3-030-24219-0_7

What You Will Learn in This Chapter

This chapter discusses the passive component of conduction that occurs between electrically coupled cells. In addition, the necessary conditions for wavefront propagation and thus normal cardiac function are introduced. Finally, this chapter will explore how intercellular coupling can be altered in pathology, and the effects this has on the electrical function of the heart.

Learning Objectives

- Understand existing theories of cardiac conduction at both a gross and molecular level
- Be able to discuss gaps in current knowledge, including the variable functional effects of phosphorylation status at serine residues
- Be able to give two examples of neoteric therapies specific to cardiac conduction

7.1 Cardiac Conduction and Gap Junctions

7

Cardiac conduction refers to the propagation of electrical activity through the heart and can be separated into active and passive components. The active component relies on membrane-bound events that are responsible for the cell's response to excitation, previously discussed in ▶ Chap. 5. This section will discuss the passive component of conduction, occurring between electrically coupled cells.

Cardiac myocytes are electrically coupled through gap junctions, which provide a conduit for charged particles and small molecules of up to 1000 Daltons in size. Gap junctions are predominantly located at the intercalated discs, where they form clusters known as *plaques* [1].

Individual gap junctions consist of many gap junctional channels, which in turn are formed by the union of two hemi-channels contained on the surface of two cells in contact with one another. The hemi-channel, or *connexon*, is in turn formed by six *connexin* proteins.

Many different connexin proteins have been identified in the human heart, each with different molecular weights and functional properties. Their nomenclature is based on their molecular weight. The most prevalent types are connexin40 (Cx40), Cx43, and Cx45. The distribution of these connexins in the mammalian heart is shown in ▫ Fig. 7.1. Broadly, gap junctions are formed by Cx43 in ventricular myocytes and Cx40 and Cx43 in atrial myocytes, with Cx45 of particular importance in the specialised conduction tissues. Due to the distribution of connexins varying between species, experimental findings in animal models may not be directly transferable to humans [2, 3].

7.1.1 Safety of Conduction

The *safety factor* (SF) for conduction describes the likelihood of successful action potential propagation across the myocardium and is determined by the relationship between two variables. The '*source*' refers to excited cells which provide charge to depolarise neighbouring unexcited cells, acting as an electrical 'sink' or load. For propagation to succeed, the source must overcome the sink, described by a SF of 1 or more. When the source provides insufficient charge to depolarise the sink, SF falls to below 1 and conduction block occurs [4].

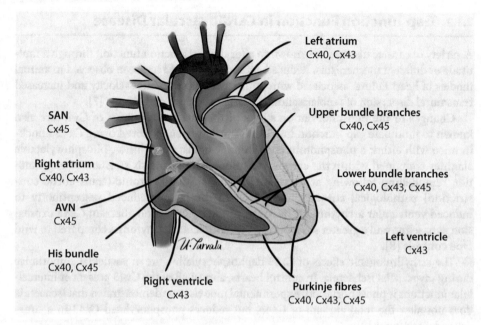

SAN
Cx45

Right atrium
Cx40, Cx43

AVN
Cx45

His bundle
Cx40, Cx45

Right ventricle
Cx43

Left atrium
Cx40, Cx43

Upper bundle branches
Cx40, Cx45

Lower bundle branches
Cx40, Cx43, Cx45

Left ventricle
Cx43

Purkinje fibres
Cx40, Cx43, Cx45

U. Tarvala

▢ **Fig. 7.1**   Diagram of the pattern of expression for the three principle connexins in the mammalian heart

▢ **Table 7.1**   Table showing safety factor for conduction and conduction velocity at different levels of inter-cellular coupling

Tissue	Strong intercellular coupling	Moderate intercellular coupling	Weak intercellular coupling
Safety factor	Moderate	High	Low
Conduction velocity	Rapid	Slow	Very slow

The importance of SF can be illustrated by modelling the propagation of an afterdepolarisation originating from a cluster of contiguous myocytes. In well-coupled tissue, this cluster of myocytes is electrically connected to many unexcited cells, which act as a large electrical sink. This means that the SF is low, and afterdepolarisations rarely propagate into the surrounding myocardium. In poorly coupled tissue, the sink is comparatively smaller and afterdepolarisations are therefore much more likely to spread, resulting in a premature ventricular complex [5].

Intercellular coupling is also an important determinant of conduction velocity, evidenced by the reduced conduction velocity observed in human and human and guinea-pig myocardium that has previously undergone the pharmacological uncoupling of gap junctions using carbenoxolone [4, 6] (▢ Table 7.1).

7.1.2 Gap Junction Function in Cardiovascular Disease

A variety of cardiac diseases are known to alter the gap junction function through a multitude of different mechanisms. Reduced expression of Cx43 has been observed in animal models of heart failure, associated with both reduced conduction velocity and increased transmural dispersion of repolarisation. These are both pro-arrhythmic [7].

Changes to the phosphorylation status of the carboxy terminus of Cx43 are also known to modulate gap junction function. This has been explored using Cx43 knock-in mice with either a phosphomimetic glutamic acids, or with non-phosphorylatable alanines contained within the carboxy terminus. The former is resistant to gap junction remodelling following acute (global ischaemia) and chronic (trans-aortic constriction) pathological stimuli, in addition to displaying reduced susceptibility to induced ventricular arrhythmias. In contrast, the latter exhibit aberrant Cx43 expression at baseline and a greater susceptibility to ventricular arrhythmia compared to wild type controls [8].

The arrhythmogenic effects of Cx43 dephosphorylation are of particular importance during myocardial ischaemia. In control hearts, almost all of the Cx43 present at intercellular junctions is phosphorylated. Experimental models have demonstrated that ischaemia does not alter the total amount of Cx43, but reduces phosphorylated Cx43 in a time-dependent manner [9].

Perturbations of connexin trafficking can also reduce intercellular coupling. Microtubules have a well-characterised role in transporting Cx43 hemi-channels to the plasma membrane where they are able to form functional gap junctions. Ischaemic human hearts have displaced microtubules-associated proteins. This limits the delivery of Cx43 to its canonical position at the intercalated disc and consequently reduces intercellular coupling [10].

Gap junction localisation may also be affected by alterations to accessory proteins. One of the most well-explored examples is zonula occludens (ZO)-1, which interacts with Cx43. Interfering with the Cx43–ZO-1 interaction leads to increased gap junction plaque size, suggesting that ZO-1 regulates the accretion of Cx43-containing gap junctions [1]. Whilst it is clear that ZO-1 possesses the ability to modulate gap junction localisation, its role in disease is less clear. Failing human hearts display a 95% reduction in ZO-1 expression, but this is accompanied by a paradoxical reduction in Cx43 staining at intercalated discs [11]. Clearly, more work is required to understand the effects of ZO-1 in health and disease.

7.1.3 Proarrhythmic Effects of Altered Gap Junction Function

Alterations to cellular coupling are closely associated with the formation of re-entrant arrhythmias, but it is important to note that re-entry may also result from numerous other functional and structural modifications of the myocardium. This form of conduction occurs when a propagating wavefront encounters an obstacle and circulates around the area to re-excite tissue at the site of origin. This rotating wavefront may proceed through several cycles, repeatedly exciting the surrounding myocardium. This is in stark contrast to normal cardiac conduction, in which a wavefront originates from the cardiac pacemaker and activation and recovery of the myocardium are completed before the arrival of the next stimulus [12].

Re-entry relies on both the conduction velocity (Θ) and refractory period (t_r). The product of these two variables defines the wavelength of excitation (λ; $\lambda = \Theta t_r$). Re-entry can occur when the length of the re-entrant circuit is greater than or equal to the wavelength of excitation.

Under these conditions, a so-called *excitable gap* is formed: excitable tissue that lies between the head of the circulating wavefront and the repolarising tail of the preceding wave. Impaired intercellular coupling reduces conduction velocity and thus λ. This renders the myocardium more susceptible to arrhythmia by reducing the minimum path length necessary for re-entry [4].

Re-entrant arrhythmias include atrial fibrillation, atrial flutter, ventricular tachycardia and ventricular fibrillation. Peters et al. were among the first to explicitly correlate gap junctional changes with re-entrant circuits [13]. Myocardial infarction was generated in six mongrel dogs by surgical ligation of the left anterior descending artery (LAD). Four days later, an electrode array was used to produce activation maps of the myocardium during induced ventricular tachycardia (VT). These maps showed two connected re-entrant circuits; one circulating clockwise and the other counter-clockwise. Both circuits travel through a central passage termed the central common pathway, giving rise to a characteristic figure-of-eight shape.

In the same study, sections of myocardium were also immunolabelled for Cx43. Myocytes in the sub-epicardial tissue overlying the infarcted region (the so-called *epicardial border zone, EBZ*) displayed Cx43 abnormally located on the lateral surface of cells. In some regions, cells with altered gap junction distribution were present throughout the entire EBZ, extending all the way to the epicardial surface. Crucially, these regions of full-thickness gap junction disarray were shown to correlate with the location of the central common pathway of circuits responsible for VT [13].

7.1.4 Functional Effects of Gap Junction Lateralisation

Gap junction lateralisation has been noted in numerous pathological conditions, but its functional effects remain unclear. In healthy myocardium, the predominant localisation of gap junctions at intercalated discs helps to establish anisotropic conduction. This refers to conduction that occurs at higher velocity in the longitudinal than transverse direction. Lateralisation would be expected to increase gap junction current travelling in a transverse direction, thus reducing anisotropy. This hypothesis has been explored in the healing canine infarct model described above. Surprisingly, there is normal transverse conductance between cells in the common central pathway with lateralised gap junctions [14].

Other groups have even found a paradoxical reduction in side-to-side coupling in these myocytes, raising the possibility that lateralised gap junctions are non-functional [15]. They may merely be a marker of disease, rather than an active participant in altered wavefront propagation. Interestingly, gap junction conductance in the longitudinal direction was normal in both of these studies, but conduction slowing through the common central pathway was present nonetheless. This has been attributed to modulation of sarcolemmal ion currents such as I_{Na}, and illustrates the importance of considering both the active and passive components of cardiac conduction [14].

7.2 What We Don't Know

7.2.1 Functional Effects of Phosphorylation Status at Serine Residues

As previously described, myocardial ischaemia is known to result in the dephosphorylation of Cx43. Whilst this statement is generally true, it fails to acknowledge the role of specific serine residues of Cx43. Some of these respond to ischaemia in the manner expected with, for example, an eight-fold reduction in the presence of Cx43 phosphorylated at serines 325/328/330 in ischaemic myocardium. The functional importance of this can be demonstrated by transfecting Cx43$^{-/-}$ mice with a mutated form of Cx43 in which these three serine residues cannot be phosphorylated. In these cells, the event frequency for fully open channels was markedly lower than in wild type controls. Moreover, they also displayed impaired transfer of Lucifer Yellow dye between cells, which is a surrogate marker of gap junction conductance [16].

In contrast, some serine residues appear to be phosphorylated in response to ischaemia. Serine 368 phosphorylation is increased following the imposition of no-flow ischaemia. Cx43 phosphorylated in this manner remains predominantly located at the intercalated disk, despite extensive lateralisation of total Cx43.

However, the functional implications of this modification remain unclear. Serine 368 phosphorylation might reduce total channel conductance and ion permeability [17]. Further complexity is added by considering the regulatory role of Cx43 residues. Phosphorylation of serine 365 is ubiquitous in homeostatic cells, however is lost following ischaemia. This is associated with more efficient phosphorylation of serine 368. It is therefore possible that phosphorylation of Cx43 at serine 365 protects cells against the effects of ischemia by limiting serine 368 phosphorylation [18].

All of these studies illustrate the complex and sometimes contrasting alterations to the phosphorylation state of Cx43 under conditions of ischaemia. The functional implications of serine residue phosphorylation in response to ischaemia are yet to be fully characterised.

7.2.2 Non-Gap Junctional Transmission of Electrical Impulses

Some of the most curious findings in the field of cardiac conduction have been made in connexin knock-out mice. It is evident that cardiac-restricted inactivation of Cx43 almost abolishes side-to-side and end-to-end gap junctional coupling between cell pairs. From these findings, it might be assumed that wavefront propagation on the whole-heart level would also be abolished. In fact, conduction velocity is reduced by just 50% in both the longitudinal and transverse directions [19].

This might be explained by the ephaptic coupling hypothesis. Conduction through this mechanism relies on close spatial apposition of adjacent cell membranes and on a high density of sarcolemmal $Na_v1.5$ channels. Under these conditions, myocyte depolarisation is hypothesised to draw Na^+ ions from the cleft between apposed cells, with the removal of positive ions making the voltage of the cleft more negative. The potential difference between the extracellular space and the neighbouring myocyte is therefore more positive. If a sufficiently positive potential differences is reached, $Na_v1.5$ channels in the neighbouring cell would open, allowing excitation to propagate from one cell to another in a gap junction-independent manner [2].

In reality, conduction is likely to occur through both gap junctional and ephaptic mechanisms, in so-called '*mixed mode*' coupling, supported by computation tissue models showing that enhanced ephaptic coupling reduces the gap junctional coupling necessary for successful conduction [20]. The mechanisms underlying cell-to-cell electrical propagation, and their relative importance, are yet to be fully elucidated.

7.3 Where We're Heading

7.3.1 Pharmacological Enhancers of Gap Junction Function

Rotigaptide (ZP-123) is an under-investigation clinical compound that has been suggested as a novel treatment for cardiac arrhythmias due to its enhancement of gap junctional coupling.

Initial experiments on this drug focused on its potential influence on the acute electrophysiological response to ischaemia, with myocardial infarction induced by surgical ligation of the left anterior descending (LAD) artery in mongrel dogs. Infusion of rotigaptide significantly reduced the incidence of induced VT in the 1–4 hours post-infarction [21]. Whilst this study demonstrated the potentially beneficial effects of rotigaptide in the context of infarction, it did not explore whether this benefit could be extended beyond the course of a few hours.

More recently, attention has turned towards the potential role of rotigaptide in influencing the pathophysiological processes that underlie infarction, meaning it may confer protection against arrhythmia that extends beyond the duration of treatment. This has been explored in rats given a loading dose of rotigaptide immediately before ligation of the LAD and for the subsequent 7 days. Rotigaptide-treated hearts had reduced arrhythmia susceptibility at 4 weeks post-infarction that could not be explained by a reduction in Cx43 lateralisation or scar size.

The anti-arrhythmic effect is most likely to result from a less heterogenous pattern of fibrosis at the infarct border zone (IBZ). Heterogeneity of scarring in this region is known to provide a substrate for re-entry, as the interdigitation of myocardium and fibrosis produces a long and tortuous path of conduction. There was also a concurrent increase in conduction velocity across the IBZ, which would reduce the wavelength of excitation and thus the occurrence of re-entry.

Gap junction enhancement is hypothesised to be the mechanism responsible for these observations. Closure of gap junctions during ischaemia prevents the cell-to-cell transfer of mediators of cell death and survival. This isolates cells within a cluster, meaning there is likely to be a greater spread of cell survival and death. Enhancement of gap junction conductance with rotigaptide would allow these mediators to be shared between surrounding cells, producing clusters of cells with a shared fate [22]. More work is required to establish whether the beneficial effects of rotigaptide in animal models can be translated to humans.

7.3.2 Modulation of Accessory Proteins

Modulation of accessory proteins is another proposed mechanism to therapeutically augment gap junction function. αCT1 is a peptide memetic of carboxyl terminus of Cx43 that has been shown to reduce the interaction between Cx43 and the accessory protein ZO-1. As discussed previously, this has been shown to increase gap junction plaque size in vitro [1].

Its effects in vivo have been explored using mice exposed to cryo-injury of the left ventricle. Whilst this model has little clinical relevance, it is proposed to generate a more reproducible injury than coronary artery ligation. Injuries were immediately covered with a patch containing αCT1 or vehicle solution. At 7–9 days post-injury, αCT1-treated hearts showed reduced susceptibility to induced arrhythmia, with an increased ventricular depolarisation rate. This was associated with a significant reduction in Cx43 lateralisation in the injury border zone of αCT1-treated hearts. Surprisingly, this may be due to altered Cx43 phosphorylation rather than modulation of the effects of ZO-1. Compared to controls, αCT1-treated hearts had significantly higher levels of Cx43 phosphorylated at serine 368 [23].

This form of Cx43 has been shown as preferentially localised to the intercalated disc following ischaemia [17] and might improve intercellular coupling to thus reduce the propensity for arrhythmias. This illustrates the anti-arrhythmic effects of this mechanism of gap junction modulation, but their clinical relevance is limited by the selected model and route of drug delivery.

7

> **Take-Home Message**
>
> - Normal cardiac conduction synchronises the activity of cardiomyocytes, giving rise to coordinated and efficient mechanical contraction.
> - Propagation of electrical activity across the myocardium is due in part to gap junctions, which facilitate the transfer of small molecules between neighbouring myocytes.
> - Gap junction changes in disease interfere with wavefront propagation, increasing the incidence of re-entry.
> - Novel therapies aim to modulate gap junction function following myocardial injury, reducing susceptibility to arrhythmia.

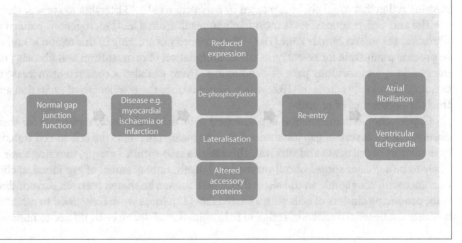

References

1. Hunter AW, Barker RJ, Zhu C, Gourdie RG (2005) Zonula occludens-1 alters connexin43 gap junction size and organization by influencing channel accretion. Mol Biol Cell 16(12):5686–5698
2. Veeraraghavan R, Poelzing S, Gourdie RG (2014) Intercellular electrical communication in the heart: a new, active role for the intercalated disk. Cell Commun Adhes 21(3):161–167

3. Severs NJ, Bruce AF, Dupont E, Rothery S (2008) Remodelling of gap junctions and connexin expression in diseased myocardium. Cardiovasc Res 80(1):9–19
4. Kléber EGK, Rudy Y (2018) Basic mechanisms of cardiac impulse propagation and associated arrhythmias [cited 2018 May 28]. Available from: https://doi.org/10.1152/physrev.00025.2003
5. Xie Y, Sato D, Garfinkel A, Qu Z, Weiss JN (2010) So little source, so much sink: requirements for after-depolarizations to propagate in tissue. Biophys J 99(5):1408–1415
6. Dhillon PS, Gray R, Kojodjojo P, Jabr R, Chowdhury R, Fry CH et al (2013) Relationship between gap-junctional conductance and conduction velocity in mammalian myocardium. Circ Arrhythm Electrophysiol 6(6):1208–1214
7. Poelzing S, Rosenbaum DS (2004) Altered connexin43 expression produces arrhythmia substrate in heart failure. Am J Phys Heart Circ Phys 287(4):H1762–H1770
8. Remo BF, Qu J, Volpicelli FM, Giovannone S, Shin D, Lader J, et al (2018) Phosphatase-resistant gap junctions inhibit pathological remodeling and prevent arrhythmias [cited 2018 May 28]. Available from: https://doi.org/10.1161/CIRCRESAHA.111.244046
9. Beardslee MA, Lerner DL, Tadros PN, Laing JG, Beyer EC, Yamada KA et al (2000) Dephosphorylation and intracellular redistribution of ventricular Connexin43 during electrical uncoupling induced by ischemia. Circ Res 87(8):656–662
10. Smyth JW, Hong T-T, Gao D, Vogan JM, Jensen BC, Fong TS et al (2010) Limited forward trafficking of connexin 43 reduces cell-cell coupling in stressed human and mouse myocardium. J Clin Invest 120(1):266–279
11. Laing JG, Saffitz JE, Steinberg TH, Yamada KA (2007) Diminished zonula occludens-1 expression in the failing human heart. Cardiovasc Pathol 16(3):159–164
12. Antzelevitch C, Burashnikov A (2011) Overview of basic mechanisms of cardiac arrhythmia. Cardiac Electrophysiol Clin 3(1):23–45
13. Peters NS, Coromilas J, Severs NJ, Wit AL (1997) Disturbed Connexin43 gap junction distribution correlates with the location of reentrant circuits in the Epicardial border zone of healing canine infarcts that cause ventricular tachycardia. Circulation 95(4):988–996
14. Cabo C, Yao J, Boyden P, Chen S, Hussain W, Duffy H et al (2006) Heterogeneous gap junction remodeling in reentrant circuits in the epicardial border zone of the healing canine infarct. Cardiovasc Res 72(2):241–249
15. Yao J-A (2003) Remodeling of gap Junctional Channel function in Epicardial border zone of healing canine infarcts. Circ Res 92(4):437–443
16. Lampe PD, Cooper CD, King TJ, Burt JM (2006) Analysis of Connexin43 phosphorylated at S325, S328 and S330 in normoxic and ischemic heart. J Cell Sci 119.(Pt 16:3435–3442
17. Ek-Vitorin JF (2006) Selectivity of connexin 43 channels is regulated through protein kinase C-dependent phosphorylation. Circ Res 98(12):1498–1505
18. Solan JL, Marquez-Rosado L, Sorgen PL, Thornton PJ, Gafken PR, Lampe PD (2007) Phosphorylation at S365 is a gatekeeper event that changes the structure of Cx43 and prevents down-regulation by PKC. J Cell Biol 179(6):1301–1309
19. Yao J-A, Gutstein DE, Liu F, Fishman GI, Wit AL (2003) Cell coupling between ventricular myocyte pairs from connexin43-deficient murine hearts. Circ Res 93(8):736–743
20. Weinberg SH (2017) Ephaptic coupling rescues conduction failure in weakly coupled cardiac tissue with voltage-gated gap junctions. Chaos 27(9):093908
21. Xing D, Kjølbye AL, Nielsen MS, Petersen JS, Harlow KW, Holstein-Rathlou N-H et al (2003) ZP123 increases gap junctional conductance and prevents reentrant ventricular tachycardia during myocardial ischemia in open chest dogs. J Cardiovasc Electrophysiol 14(5):510–520
22. Ng FS, Kalindjian JM, Cooper SA, Chowdhury RA, Patel PM, Dupont E et al (2016) Enhancement of gap junction function during acute myocardial infarction modifies healing and reduces late ventricular arrhythmia susceptibility. JACC Clin Electrophysiol 2(5):574–582
23. O'Quinn MP, Palatinus JA, Harris BS, Hewett KW, Gourdie RG (2011) A peptide mimetic of the Connexin43 carboxyl terminus reduces gap junction remodeling and induced arrhythmia following ventricular injury. Circ Res 108(6):704–715

Cell-Based Tachyarrhythmias and Bradyarrhythmias

Rohin K. Reddy, Ben N. Cullen, Ahran D. Arnold, and Zachary I. Whinnett

© Springer Nature Switzerland AG 2019
C. Terracciano, S. Guymer (eds.), *Heart of the Matter*, Learning Materials in Biosciences,
https://doi.org/10.1007/978-3-030-24219-0_8

What You Will Learn in This Chapter

Arrhythmias can have immediate clinical consequences because effective cardiac output is determined by efficient electromechanical coupling of the myocardium and coordination between and within cardiac chambers. This chapter will discuss specific bradyarrhythmias and tachyarrhythmias, providing an overview of their pathophysiology, clinical characteristics and treatment options. As the heart constitutes a functional syncytium, the cellular basis of arrhythmia will first be discussed before considering its effects across the larger myocardial mass as well as clinical manifestations.

Learning Objectives

— Discuss early and delayed afterdepolarisations, including the relevant ionic currents involved.
— Appreciate how disturbances of impulse formation and conduction can lead to arrhythmia.
— Provide an overview of both supraventricular and ventricular tachycardias, including treatment.

8

8.1 An Introduction to Cardiac Arrhythmias

The cardiac action potential results from a tightly orchestrated process of ion channel opening and closing. This can be modulated by the sympathetic and parasympathetic nervous systems, which operate antagonistically to allow continual and rapid adaptation to varying demands placed upon the body. *Cardiac arrhythmia* is a broad term which refers to a multitude of electrical abnormalities within the myocardium. The clinical presentation of cardiac arrhythmias is also variable, which ranges from no apparent symptoms to disabling symptoms, heart failure, sudden cardiac death (SCD) or the consequences of systemic thromboembolism. SCD, in particular, is a significant public health issue, and is estimated to cause 300,000–375,000 annual deaths in the United States, which represents around half of all cardiovascular deaths [1]. Indeed, with around 50% of SCD cases representing the first expression of disease [2], it is vital to identify those at elevated risk in order to expediently implement appropriate management.

In 1887 at St Mary's Hospital, London, August Waller recorded the first human electrocardiogram (ECG), becoming the first to characterise the electrical activity of the heart [3]. However, it was Willem Einthoven, having adapted the work of Waller, who published the initial observations of normal and abnormal electrocardiograms in 1906, thereby being the first to demonstrate the ECG's capacity to diagnose a range of cardiac abnormalities [4]. Despite the technological limitations at that time, Einthoven's ECG was able to identify abnormalities such as atrial and ventricular hypertrophy, premature ventricular complexes, and atrial flutter, among others. However, it was not until the first intracardiac recording of His bundle potentials was made in 1969 that the field of clinical cardiac electrophysiology began in earnest [5].

It was these early studies that precipitated the development of multielectrode catheters, introduced in the 1970s for the intraoperative mapping of Wolff-Parkinson-White syndrome and ventricular tachycardia [6]. From this early therapeutic utilisation of electrophysiology, the following decades were characterised by multiple paradigm-shifting advances including catheter ablation, permanent pacemakers, implantable cardioverter-defibrillators (ICDs) and three-dimensional electro-anatomical mapping. Thus, whilst it is still a relatively new field, the rapid growth of cardiac electrophysiology has catalysed the successful diagnosis and management of the arrhythmias introduced within this chapter.

8.2 Mechanisms of Arrhythmia

8.2.1 Tachyarrhythmia

Normal electrical impulses within the heart are initiated from the sinoatrial node (SAN): a collection of specialised myocytes localised in the high right atrium that exhibit automaticity, the capacity to spontaneously initiate action potentials. However, action potentials may also initiate from regions of myocardium outside of the SAN, known as *ectopic foci*, in a phenomenon termed *abnormal automaticity* [7]. These foci occur within the myocardium following ischaemic insults, sympathetic nervous over-stimulation, electrolyte abnormalities and use of certain drugs; their spontaneous electrical activity can result in the capture of normal pacemaker impulses originating from the SAN [7]. Moreover, ectopic foci are associated with triggered arrhythmias through the spontaneous depolarisation of cardiac myocytes and subsequent activation of voltage-gated Na^+ or Ca^{2+} channels, resulting in the generation of additional (non-SAN-derived) action potentials [7].

When spontaneous depolarisation occurs during the plateau or late phase of repolarisation, it is termed an *early afterdepolarisation* (EAD). Conversely, if occurring during the diastolic period, it is termed a *delayed afterdepolarisation* (DAD). Both EADs and DADs are due to abnormal intracellular Ca^{2+} handling.

8.2.2 Early Afterdepolarisations (EADs)

EADs derive from a prolongation of the action potential duration, bradycardia and/or a reduction in the magnitude of repolarising currents [8]. As K^+ is the primary ion effluxed from cardiomyocytes during repolarisation, it is logical that the following factors are associated with the diminution of repolarising currents, and thus EADs [9]:

- Hypokalaemia, decreasing I_{K1} magnitude
- Drugs that result in K^+ channel blockade
- Channelopathies such as Long QT syndrome

Due to a reduction in the magnitude of repolarising currents, abnormal increases in the net inward current can overcome repolarisation, promoting the generation of small depolarisations in the voltage range -20 mV to $+10$ mV that typically occur near the end of the plateau phase (phase 3) of the action potential [8]. Notably, this is the voltage range in which $I_{Ca,L}$ activates, creating a region of overlap between the inactivation and activation range of $I_{Ca,L}$. Within this voltage range, termed the *window current*, a proportion of the Ca^{2+} channels previously stimulated recovers and is able to be reactivated [10].

If the aforementioned net inward current is sufficient to overcome repolarisation, the small depolarisations can mediate the triggering of enhanced Ca^{2+} loads from the sarcoplasmic reticulum [10, 11]. This results in aftercontractions, in addition to greater amounts of inward current being carried by I_{NCX}, magnifying the EAD size. Thus, whilst I_{CaL} is the principal EAD driver, the aberrant handling of calcium also plays a critical role through the enhancement of I_{NCX}.

8.2.3 Delayed Afterdepolarisations (DADs)

DADs are usually associated with phase 4 of the cardiac action potential (AP): the transitional stage between the completion of the preceding action potential and initiation of the current pacemaker potential in the SAN. DADs are commonly observed in high intracellular Ca^{2+} loads, whether in the cytoplasm or sarcoplasmic reticulum (SR).

This phenomenon classically occurs in states of tachycardia or following the administration of β-adrenergic receptor agonists or cardiotonic steroids (e.g., digoxin) [12, 13]. Elevated calcium within the SR increases the likelihood of spontaneous Ca^{2+} release into the cytoplasm after repolarisation, resulting in its removal by the Na^+/Ca^{2+} exchanger (NCX). As previously described, this transfer generates a net inward current of +1, potentially generating depolarisations of sufficient charge to activate Na^+ channels and initiate a premature upstroke in membrane potential. Clinically, DADs are also associated with myocardial infarction (MI), both due to the partially depolarised state of infarcted tissue and the characteristic accompanying catecholamine surge [13, 14] (◙ Fig. 8.1).

8.2.4 Tachyarrhythmia

Re-entry can arise within microscopic areas of the myocardium, or more globally, affecting the entire cardiac chamber. Two conditions must be met for the commencement of re-entry [15]:
1. *Unidirectional Block*: a zone of myocardium which acts like a diode, only permitting the transmission of electrical impulses in one direction.
2. *Slow Conduction Speed*: such that the same tissue region is no longer refractory by the time impulses reach the distal end of the unidirectional block.

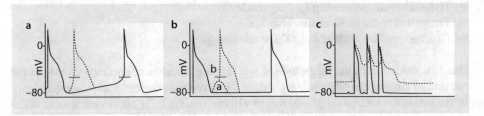

◙ **Fig. 8.1** Cellular mechanisms of abnormal impulse formation. **a** Accelerated normal automaticity in a spontaneously depolarising pacemaker cell. **b** DAD in a ventricular cell, typically after a repolarisation (dashed line, **a**) and manifesting as abnormal "hump-like" depolarisations. If sufficient to meet the electrical threshold, they can cause abnormal extra beats (dashed line, **b**). **c** Early after-depolarizations. Action potential traces in two cardiomyocytes. The upper trace demonstrates a failure to repolarise, generating repeated early afterdepolarisations that cause an adjacent cell to fire repeatedly, as seen in the lower trace. (Permission to reuse image from [14])

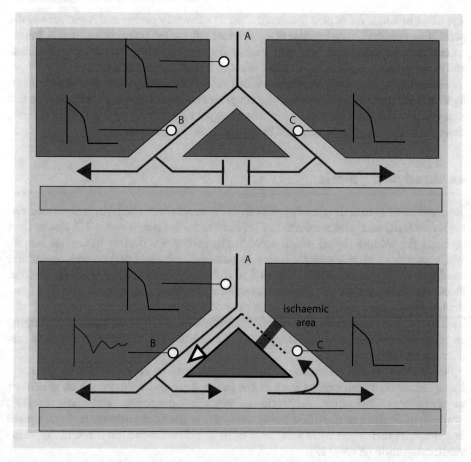

◘ Fig. 8.2 Schematic representation of re-entrant circuit formation. (Adapted from [16])

A diagram of a typical re-entrant circuit is shown below [16]. ◘ Figure 8.2a shows a region of cardiac tissue exhibiting branching. Impulses proceed normally down the main branch, A, before bifurcating to transmit down branches B and C. Finally, a similar bifurcation occurs in each branch once again. As shown, if the two APs impact, they 'cancel each other out' so to speak, due to the establishment of refractory conditions in the membrane. This means that the first impulse to meet the tissue makes the tissue refractory, so the second impulse cannot activate it and therefore cannot conduct beyond that tissue. Equally, the first impulse cannot activate the tissue already activated by the second impulse, thus there is no activation in either direction.

However, if the myocardial tissue is diseased, it may only facilitate conduction in one direction, i.e. a unidirectional block is present. This could be due to a number of factors; however, it often originates from ischaemia-induced gap junction remodelling, explained in previous chapters. This phenomenon is illustrated in ◘ Fig. 8.2, whereby at bifurcation point A, impulses are now exclusively transmitted down branch B as a unidirectional block is present in the shaded 'ischaemic area' in C [16].

An action potential is recorded later at C however, because the original impulse is able to move back up branch C towards A, retrogradely traversing the unidirectional block.

The tissue just after A has now recovered from its refractory period and is excitable again, permitting re-entry of the impulse at the original bifurcation point. Subsequently, the re-entrant impulse can now travel down B again, depolarising the tissue earlier than normal and in the process, establishing a circuit of much higher frequency impulses [16].

Following MI, the heart comprises regions of functional myocardium in close proximity to areas of dysfunctional intercellular coupling and zones of reparative fibrosis. In this way, the post-infarct heart represents an archetypal substrate for re-entrant arrhythmia, characterised by intratissue variation in both excitability and conductivity.

8.3 Bradyarrhythmia

The heart's ability to maintain adequate cardiac output is determined by two parameters: heart rate (HR) and stroke volume (SV). Bradycardia is defined as any HR slower than 60 beats per minute (bpm) which accordingly, mediates a decline in cardiac output. When pathological, bradycardia is due to sinus node dysfunction or atrioventricular conduction abnormalities [17]. It is vital to distinguish pathological bradycardia from its physiological counterpart, which is commonly encountered in resting athletes (particularly endurance athletes) and is generally considered benign. These so-called 'athlete's hearts' have undergone hypertrophic adaptations to repetitive loading, facilitating an increase in SV and compensatory reduction in basal HR to maintain the same cardiac output [18].

In contrast, sustained pathological bradycardia can be characterised by a chronic hypoperfusion of vital organs such as the brain, liver and kidneys, which manifests as symptoms including palpitations, syncope, dyspnoea and fatigue [17]. Bradyarrhythmias are classified based on whether they originate from disturbances in a) impulse formation within the SAN or b) impulse conduction through the atrioventricular node (AVN) [17].

8.4 Disturbance of Impulse Formation

Under normal conditions, the SAN initiates all cardiac impulses, with both intrinsic and extrinsic factors contributing to its dysfunction. These include:

Intrinsic Commonly caused by idiopathic SAN dysfunction, with around 30% of cases instead deriving from vascular occlusion of the node secondary to ischaemic heart disease. Moreover, dysfunction has been demonstrated as occurring due to hypertensive remodelling, or remodelling instigated by infiltrative conditions such as sarcoidosis, distorting the nodal cell membrane [19].

Arrhythmias such as atrial flutter and atrial fibrillation (AF) have also been associated with nodal dysfunction via a complex remodelling process.

Extrinsic Drug-related disturbances represent by far the commonest cause of extrinsic SAN dysfunction, with mainstay therapies such as β-blockers and Ca^{2+} channel blockers among the key players, in addition to digoxin [20]. Less common extrinsic causes include increased vagal tone and electrolyte derangement, both of which can also reduce the nodal firing rate.

8.4.1 Insufficiency of SAN Impulse Generation

Briefly, insufficiency of SAN impulse generation can be broadly categorised into five variants:

1. *Sinus Arrest:* a failure of the SAN to discharge pacemaker potentials, manifesting as the absence of atrial depolarisation and periods of ventricular asystole [21]. This may be episodic, such as in vasovagal syncope and carotid hypersensitivity.
2. *Exit Block:* failed propagation of pacemaker potentials upon leaving the SAN, which normally generates impulses. This is classically represented on ECG as an interval between P waves that is a multiple of the normal interval [21]. Exit block is differentiated from sinus arrest by the fact that sinus arrest randomly restarts, whilst the exit block interval may be predicted.
3. *Sinus Bradycardia:* characterised by a slow rate of sinus node discharge. If patients are symptomatic and the bradycardia is both persistent and irreversible, the patient may require a pacemaker. However, as previously mentioned, it is vital to distinguish this from resting bradycardia.
4. *Chronotropic Incompetence:* typified by an inability to alter the heart rate in response to varying metabolic demand. This results in severe, symptomatic exercise intolerance, with patients often requiring a pacemaker.
5. *Tachycardia-Bradycardia Syndrome:* intermittent episodes of slow and fast sinus rates, representing a common indication for cardiac pacing [21].

8.4.2 Disturbance of Impulse Conduction

Pacemaker potentials generated in the SAN are transmitted to the AVN via the anterior, middle, and posterior internodal tracts, including Bachmann's bundle of the anterior tract. These broadly-traversing pathways provide electrical stimulation to the atrial myocardium, generating the P wave visible on an ECG. The AVN is located at the centre of Koch's triangle, and provides the solitary AV electrical connection through the central fibrous node of the annulus fibrosus. Transmission is decelerated at the AVN to allow completion of active ventricular filling, corresponding to the PR interval on the ECG. Many disorders of impulse conduction are associated with AVN dysfunction, often overlapping with causes of SAN dysfunction.

Whilst the AV block can occasionally be congenital in nature, most disturbances are acquired, often secondary to the usage of drugs such as amiodarone and β-blockers [22]. An MI affecting the AV nodal tissue should also be considered, as should degenerative changes such as calcification and fibrosis, the latter of which can follow from infarction. Infections including Lyme and Chagas disease are also important, albeit rarer causes of dysfunction, as are manifestations of infection such as myocarditis and endocarditis [23]. As with sinus insufficiency, AV blocks can be broadly categorised into major variants:

- *First Degree AV Block:* impulse passes the AVN slower than usual and is defined by a PR interval >200 ms on ECG. It is usually asymptomatic.
- *Second Degree AV Block:* one or more (but not all) of the cardiac action potentials does not traverse the AV node [24]. This can be further divided into two categories:
  - *Mobitz I:* characterised by a progressive prolongation of the PR interval until there is failure to conduct and a ventricular beat is dropped [24]. This process then repeats.

— *Mobitz II:* usually concentrated in the His-Purkinje system, involving regularly-dropped ventricular beats with no change in the PR interval prior to transient failure of conduction [25]. Mobitz II is more severe than Mobitz I and may progress to complete heart block and potential cardiac arrest [24].

— *Third Degree AV Block:* also referred to as *complete heart block*. There is no impulse conduction between the atria and ventricles, with complete AV dissociation. Consequently, an accessory pacemaker in the ventricular myocardium is needed for the generation of a so-called *escape rhythm*, independent of the SAN [26]. This separation of electrical activity subsequently produces two discrete ECG rhythms, with a lack of synchronicity between the P waves (atrial depolarisation) and QRS complexes (ventricular depolarisation) [26].

Blocks in cardiac conduction pathways often occur either in the bundle branches that course along the interventricular septum, or near the AVN, which impairs conduction between the atria and ventricles [27, 28]. Abnormal conduction through the left or right bundle branches serves to prolong the depolarisation of specific ventricular tissue zones, observable as a broadening of the QRS complex on ECG [28]. The main causes of bundle branch block are idiopathic, aortic stenosis, coronary artery disease and MI.

8.5 Tachyarrhythmias

Tachyarrhythmia, also known as tachycardia, refers to a resting heart rate >100 bpm and can represent either a physiological response to external stimuli or a manifestation of an underlying pathology [29]. Just as bradycardic rhythms can result in decreased cardiac output and inadequate perfusion, an acceleration of the cardiac cycle (seen in tachyarrhythmia) depletes the time for ventricular filling prior to contraction, reducing the stroke volume and subsequently, cardiac output [30]. Symptoms associated with tachycardia include palpitations, chest pain, dyspnoea, and rarely, syncope [30]. Tachycardia may result from either disorders of impulse formation, or the generation of re-entrant circuits similar to those discussed earlier. Tachyarrhythmias can be considered as *supraventricular* (*supra*- meaning 'above'), affecting the atrial or AV nodal tissue, or *ventricular*. Alternatively, they can be characterised based upon their ECG appearance, as illustrated in ◘ Table 8.1.

8.6 Atrial Tachycardia

Atrial tachycardias can be due to either focal or re-entrant mechanisms, the latter of which is illustrated in ◘ Fig. 8.2. Focal atrial tachycardia, shown in ◘ Fig. 8.3a, is a rapid (usually 140–220 bpm), regular rhythm that arises from a point source of activation within the atrial myocardium [30]. This focus generates extrasystoles that superimpose on the normal atrial systole to produce inefficient, spastic contractions.

In comparison, ◘ Fig. 8.3b exemplifies atrial flutter: an example of a macro re-entrant circuit that can exist within the atria, often along the cavo-tricuspid isthmus of the right atrium. In typical atrial flutter, a single premature atrial contraction results in the establishment of a self-perpetuating re-entrant circuit that moves counter-clockwise around the tricuspid valve, propagated by differences in the refractory periods of surrounding myocardium [30]. It should be noted that in Figure 8.3 although we depict focal atrial

■ Table 8.1	Classification of tachycardia based on ECG characteristics	
	Narrow QRS	Wide QRS
Regular rhythm	AT AVNRT AVRT	SVT with BBB VT
Irregular rhythm	AF	AF with BBB VF TdP

Abbreviations: *AF* atrial fibrillation, *AT* atrial tachycardia, *AVNRT* atrioventricular nodal re-entrant tachycardia, *AVRT* atrioventricular re-entrant tachycardia, *BBB* bundle branch block, *SVT* supraventricular tachycardia, *TdP* Torsades de Pointes, *VF* ventricular fibrillation, *VT* ventricular tachycardia

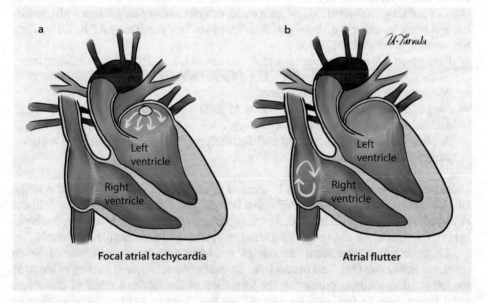

■ Fig. 8.3   a Focal atrial tachycardia. b Atrial flutter

tachycardia emanating from a left-sided focus and a right-sided macro re-entry atrial tachycardia, right-sided focal atrial tachycardias and left-sided macro re-entrant circuits are both common.

This localised re-entrant circuit of atrial flutter manifests as distinctive 'flutter waves' on ECG, usually of around 300 waves per minute (also referred to as beats per minute (bpm)) [30]. The regularity of atrial flutter also allows the ventricular rate to be defined by the atrioventricular conduction ratio. This is because of a relatively stable ratio between the number of P waves (atrial depolarisation) and QRS complexes (ventricular depolarisation), usually 2:1, which produces a ventricular rate of around 150 bpm. 'Flutter waves' give rise to a characteristic 'sawtooth' pattern on ECG. Despite this apparent regularity, atrial flutter is unstable by nature, often precipitating the development of AF. Atypical flutter is less commonly seen; however, it can exist within the left or right atrium of patients with structural heart disease, or in those who have undergone ablation procedures for AF. Atypical flutter follows a different transmission pathway and is often faster (~400 bpm).

AF is an atrial tachyarrhythmia characterised by predominantly uncoordinated atrial activation, loss of coordinated 'atrial kick' and consequent deterioration of atrial mechanical function [31]. It is the commonest cardiac arrhythmia, with a prevalence that roughly doubles with each decade of life, from 0.5% at ages 50–59 to almost 9% at ages 80–90 [31]. Common symptoms include palpitations, dyspnoea and chest pain. While there is no consensus definition of AF, classical ECG findings include absent P waves, an irregularly irregular rhythm (R-R interval does not remain constant), 'fibrillatory' waves that vary in amplitude and morphology and QRS complexes that are usually <120 ms [30]. The lack of clear P waves are due to uncoordinated, fibrillatory contractions of the atrial myocardium [30].

Importantly, AF is not a distinct arrhythmic disease: rather, it is an atrial myopathy that represents a manifestation of an underlying pathology. It is often caused by pre-existing conditions, whether cardiac such as ischaemic heart disease, rheumatic heart disease and hypertension, or non-cardiac including acute infection and electrolyte abnormalities and lung carcinoma [32]. Moreover, AF exhibits a strong association with increasing age, heart failure and diabetes mellitus. There are four patterns of AF based on their time span of presentation [32, 33]:

- *Paroxysmal AF*: self-terminating, usually within 48 hours. This subtype exhibits intermittent periods of fast palpitations. Cut-off point is 7 days.
- *Persistent AF*: lasts longer than 7 days.
- *Long-Standing Persistent AF*: continuous AF lasting for at least 1 year when it is decided to adopt a rhythm control strategy.
- *Persistent AF*: accepted by patient and doctor. Rhythm control interventions are not pursued.

These patterns were first elucidated in seminal goat studies in 1995 and gave rise to the current consensus that 'AF begets AF' – that is to say that any given period of AF initiates electrical and coupling alterations that predispose to further AF [33]. Moreover, this highlights the nature of AF as a progressive atrial myopathy, and not a simple arrhythmia.

AF represents a significant elevator of mortality risk, largely precipitated by an increased stroke risk [34]. This occurs from the stasis and subsequent pooling of blood in the left atrial appendage, promoting the formation of thrombi as a result of discordant atrial contraction. In fact, patients with AF are five times more likely to experience an ischaemic stroke than age-matched controls, with 15–20% of ischaemic stroke patients also possessing underlying AF [34]. It is for this reason that appropriate anticoagulation therapy plays a critical role in the management of AF, with the emergence of direct oral anti-coagulation agents largely superseding warfarin as the gold-standard pharmacological option [34]. Notwithstanding the success of these therapies, stroke remains a common first presentation of AF, highlighting the critical role that screening plays in facilitating early detection.

8.6.1 AVRT and AVNRT

Two notable SVTs (supraventricular tachycardia) are atrioventricular nodal re-entry tachycardia (AVNRT) and atrioventricular re-entrant tachycardia (AVRT), illustrated in ◘ Figs. 8.4 and 8.5, respectively. AVNRT is the commonest regular SVT, occurring within

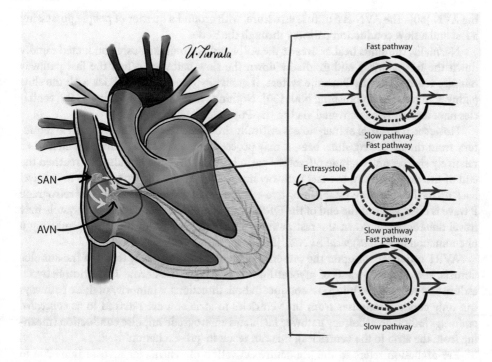

Fig. 8.4 Mechanism for atrioventricular nodal re-entrant tachycardia

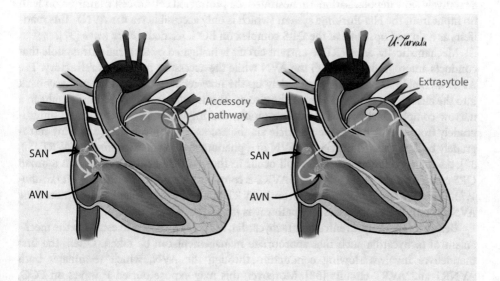

Fig. 8.5 Mechanism for atrioventricular re-entrant tachycardia

the AVN [30]. The AVN is a diffuse structure, with around a quarter of people possessing a fast and a slow conduction pathway through the node.

Normally, the sinus beat arrives at the AVN and is simultaneously conducted rapidly down the fast pathway and gradually down the slow pathway. When the fast pathway impulse reaches the His-Purkinje system, it gradually returns and collides with the slow pathway impulse, extinguishing both [30]. Nothing results from this, because the ventricles have already been activated via the His-Purkinje system.

However, if an atrial extrasystole is critically timed such that the fast pathway is refractory from the preceding sinus beat, it may proceed down the slow pathway, which has a relatively shorter anterograde refractory period. By the time the impulse has reached the end of the slow pathway, the fast pathway has recovered and the extrasystole may travel back up the fast pathway, establishing a re-entrant circuit within the AVN. This retrograde P wave is observable at the end of the QRS complex on ECG. Rarely, the extrasystole may travel anterogradely down the fast pathway and retrogradely up the slow, producing a phenomenon termed *atypical AVNRT* [35].

AVRT is dependent upon the presence of an accessory pathway through the annulus fibrosis, essentially a strand of myocardium that bridges the fibrous insulation between atrium and ventricle. While most conduct in both directions, a minority of these pathways can only conduct impulses from the ventricles to atria and are referred to as *concealed pathways* [36]. If an accessory pathway facilitates anterograde impulse conduction (meaning from the atria to the ventricles), this can result in 'pre-excitation'.

Pre-excitation refers to the premature activation of ventricular myocardium due to impulses bypassing the AV node via an anterograde accessory pathway [30, 36]. This is possible due to the propagation delay at the AVN, a feature not occurring in the accessory pathway. Importantly, if the ventricles are stimulated by an accessory pathway, they exhibit a relatively slow depolarisation, orchestrated via protracted cell-to-cell transmission with no input from the His-Purkinje system (which is only accessible via the AVN). This manifests as a slurred upstroke of the QRS complex on ECG, termed a *delta wave* [37].

Mechanistically, an AVRT re-entrant circuit is instigated by an atrial extrasystole that conducts anterogradely through the AVN while the accessory pathway is refractory. The AP is then able to transmit retrogradely up the now-recovered accessory pathway back into the atrial myocardium, forming the circuit. This is *orthodromic AVRT* and results in a narrow-complex tachycardia. Less commonly, the impulse may be conducted anterogradely from the atrium to the ventricle via the accessory pathway, before moving retrogradely back to the atrium via the AVN in a phenomenon termed *antidromic AVRT* [30, 37]. As cell-to-cell depolarisation will occur in the ventricles, this will result in a broad QRS complex with a delta wave [37]. AVRT is commonly associated with *Wolff-Parkinson-White (WPW) syndrome*, an orthodromic AVRT that leads to repeated stimulation of the AVN [37]. In WPW, the accessory pathway is termed the *bundle of Kent*.

Before treating supraventricular tachycardia, the primary aim is to establish the mechanism of tachycardia such that appropriate management can be taken. Often, the first manoeuvre involves slowing conduction through the AVN, which terminates both AVNRT and AVRT circuits [38]. Moreover, this may expose buried P waves on ECG, elucidating the sawtooth pattern typical of atrial flutter. Focal atrial tachycardia will show no response.

Slowing of conduction through the AVN may be achieved with vagal stimulation via carotid sinus massage or the Valsalva manoeuvre [38]. If these methods prove unsuccessful, the next logical step and first-line pharmacological option is adenosine: a purine nucleoside that transiently slows conduction through the AVN [38]. It is worth noting that medication rarely suppresses arrhythmias however, and therefore is often utilised as a way to minimise symptoms before catheter ablation can be scheduled, or before treatment for those whom the procedure carries unacceptably high risk. The application of radiofrequency energy via catheter ablation represents definitive management for SVTs [38]. In focal atrial tachycardia, the ectopic focus is ablated, whilst a linear lesion is created along the cavo-tricuspid isthmus in atrial flutter. Comparatively, the AVN slow pathway and accessory pathway are targeted in AVNRT and AVRT, respectively.

8.7 Ventricular Tachycardia

Ventricular tachycardia (VT) is a potentially life-threatening, macro re-entrant regular arrhythmia commonly secondary to scarring resulting from ischaemic heart disease [39]. It manifests as a broad QRS complex on ECG and is noted by its propensity to degenerate into ventricular fibrillation and asystole thereafter, preceding around 84% of SCD events [40]. VT is diagnosed by an ECG demonstrating a regular rate of >120 bpm with at least three wide QRS complexes in a row. Symptoms of VT include palpitations, dizziness and chest pain [39].

Up to 80% of VT cases ultimately derive from coronary artery disease, with cardiac ischaemia precipitating the deposition of fibrosis that later serves as a substrate for arrhythmia [39]. A significant minority of VT is related to cardiomyopathy, the majority of which are due to hypertrophic cardiomyopathy, in addition to lesser causes including primary electrical or genetic channelopathies, valvular heart disease and congenital heart disease [39]. When assessing patients, investigation tends to guide treatment, with ECG being the main confirmation of VT. The ejection fraction from echocardiography is then used to stratify risk of SCD, although it is worth noting that cardiovascular MRI is increasingly being used for this purpose. It is then important to use angiography to assess the need for treatment of any underlying coronary disease.

While anti-arrhythmic drugs were initially considered an effective treatment for VT, the 1991 CAST study demonstrated that two widely-used anti-arrhythmic compounds (encainide and flecainide) were associated with an increased mortality risk, derived from pro-arrhythmic effects [41].

Consequently, most VT patients with an elevated SCD risk are treated via an ICD [42]. Anti-tachycardia pacing within the devices utilises critically-timed extrastimuli to painlessly terminate re-entrant circuits. This is in contrast to older models, which over-relied on the delivery of painful shocks, and has made newer ICDs a mainstay of secondary SCD prevention [42]. One significant drawback of ICDs is their potential delivery of unnecessary shocks, which in addition to being acutely painful, often cause lasting psychological damage. VT ablation has been shown to reduce the incidence of VT in at-risk patients; however, it has not been demonstrated as consistently efficacious [43]. That said, cardioversion is still a recommended therapy for patients with haemodynamically unstable VT [42].

Despite the mortality increase observed with encainide and flecainide, severe yet haemodynamically-stable VT patients are acutely managed via the administration of amiodarone: a class III anti-arrhythmic agent that prolongs phase 3 (repolarisation) of the cardiac action potential to prolong the refractory period and thus delay the heart rate in several conductive tissues [44]. Importantly, due to its action of QT interval prolongation, amiodarone is contraindicated in (and has the capacity to induce) polymorphic VTs such as Torsades de Pointes (TdP) or twisting of the points. TdP is characterised by a slower repolarisation phase, especially in M cells in the mid-myocardium [44, 45]. This has two unfortunate potential consequences:

1. Prolongation of repolarisation promotes EAD development.
2. The concentration of prolongation in M cells creates an environment of variable refractory periods, increasing the risk of re-entry.

TdP is distinguishable on ECG by QRS complexes that 'twist' around the isoelectric baseline, which can result in 'R on T phenomenon', followed by circular patterns of ventricular depolarisation [45]. TdP is especially dangerous because it is associated with reductions in arterial blood pressure and syncope, often degenerating into ventricular fibrillation.

With the exception of β-blockers, currently available anti-arrhythmic agents have not been demonstrated in randomised clinical trials as effective in the primary management of patients with chronic life-threatening ventricular arrhythmias [46]. The chronic management of VT therefore primarily revolves around the use of ICDs to prevent SCD and β-blockers to treat clinical features [42].

Amiodarone is used with caution due to its pro-arrhythmic properties and broad side-effect profile, which includes both hyperthyroidism and hypothyroidism, in addition to pulmonary and hepatic toxicity [42]. A key situation when anti-arrhythmic drugs are the exclusive treatment used is in patients with structural heart disease considered to be at low risk of SCD, and in focal or structurally normal VT. Moreover, they may be used occasionally in the treatment of VT in hypertrophic cardiomyopathy or arrhythmogenic right ventricular cardiomyopathy when ICD implantation is not feasible.

A final aspect of VT care is the implementation of risk-reducing interventions for coronary disease, as ischaemic heart disease underlies most causes of VT/SCD. Measures taken should combine the treatment of hypertension, smoking cessation, reducing obesity, improving hyperlipidaemia, and improving diet and exercise [42].

8.8   What We Don't Know: Mechanisms of Atrial Fibrillation

The mechanisms of AF are complicated and the specific signalling involved is highly debatable. The original theory stemmed from the work of Moe et al., who postulated the 'multiple wavelet hypothesis', namely that a combination of multiple simultaneous re-entrant circuits within a suitable atrial substrate result in the continual excitation of atrial myocardium, precipitated by distinct initiating and sustaining factors [47]. However, in 1998, Haïssaguerre et al. proposed an alternative focal mechanism whereby electrical discharges from the pulmonary veins are responsible for initiation of the fibrillatory activity [48]. Maintaining this aberrant activity relies on the establishment of re-entrant wavefronts which 'break' as a result of heterogenous atrial structure and refractoriness, forming multiple daughter wavelets [48]. From these two influential studies and others, two major mechanistic proposals were refined:

- *Focal AF*: a rapidly firing focus (often in the pulmonary veins) cannot be conducted uniformly by all areas of the atria, leading to the wavefronts progressively 'breaking' as they travel further from the focus, leading to the occurrence of AF in distal atrial regions.

— *Substrate AF*: there is an altered substrate with such variation in conduction properties that multiple wavelets can be maintained independent of any continuous source.

With focal AF initiated as a single wavefront by a rapidly firing focus in a potentially normal tissue substrate, it can occur in relatively normal hearts, as demonstrated by Haïssaguerre. Therefore, the focal model is widely considered to be the underlying mechanism of paroxysmal AF. In contrast, substrate AF relies on a remodelled tissue substrate, aligning it more closely with the permanent AF typical of hypertension patients. That said, the natural progression of paroxysmal AF to permanent AF is well-established, which would suggest that both mechanisms are equally valid.

Underlying the progression from paroxysmal AF to permanent AF is a series of changes in ion channel expression and conduction properties that occur as the atria fibrillate, shortening their refractory period. This remodelling promotes the sustenance of fibrillatory wavefronts in an initially non-diseased substrate, explaining the observation that 'AF begets AF'. Therefore, it is likely that both initiation models explain the mechanism of AF, possibly representing different points on the same spectrum.

Take-Home Message

- A spontaneous depolarisation arising in the plateau or late phase of repolarisation is termed an early afterdepolarisation. If occurring during the diastolic period, it is referred to as a delayed afterdepolarisation.
- Disorders of impulse conduction are usually due to either a) blocks in conduction pathways or b) re-entrant circuits.
- Two criteria exist for the initiation of re-entry: 1) unidirectional block and 2) slow conduction.
- Initial atrial fibrillation causes ionic remodelling. This promotes the sustenance of fibrillatory wavefronts.

References

1. Konety SH, Koene RJ, Norby FL, Wilsdon T, Alonso A, Siscovick D et al (2016) Echocardiographic predictors of sudden cardiac death: the atherosclerosis risk in communities study and cardiovascular health study. Circ Cardiovasc Imaging 9(8):e004431
2. Goldberger JJ, Buxton AE, Cain M, Costantini O, Exner DV, Knight BP et al (2011) Risk stratification for arrhythmic sudden cardiac death: identifying the roadblocks. Circulation 123:2423
3. Waller AD (1887) A demonstration on man of electromotive changes accompanying the Heart's beat. J Physiol 8:229
4. Barold SS (2003) Willem Einthoven and the birth of clinical electrocardiography a hundred years ago. Card Electrophysiol Rev 7(1):99–104
5. Scherlag BJ, Lau SH, Helfant RH, Berkowitz WD, Stein E, Damato AN (1969) Catheter technique for recording his bundle activity in man. Circulation 39:13
6. Lüderitz B (2009) Historical perspectives of cardiac electrophysiology. Hell J Cardiol 50(1):3–16
7. Tse G (2016) Mechanisms of cardiac arrhythmias. J Arrhythmia 32:75
8. Qu Z, Xie LH, Olcese R, Karagueuzian HS, Chen PS, Garfinkel A et al (2013) Early afterdepolarizations in cardiac myocytes: beyond reduced repolarization reserve. Cardiovasc Res 99:6
9. Liu GX, Choi BR, Ziv O, Li W, de Lange E, Qu Z et al (2012) Differential conditions for early afterdepolarizations and triggered activity in cardiomyocytes derived from transgenic LQT1 and LQT2 rabbits. J Physiol 590:1171

10. Zeng J, Rudy Y (1995) Early afterdepolarizations in cardiac myocytes: mechanism and rate dependence. Biophys J 68:949
11. Viswanathan PC, Rudy Y (1999) Pause induced early afterdepolarizations in the long QT syndrome: a simulation study. Cardiovasc Res 42:530
12. Xie JT, Lowell TK, Yuan CS (2000) Extracellular detection of delayed afterdepolarization of cardiac fibers using signal averaging technique. Acta Pharmacol Sin 21(11):977–985
13. BOUTJDIR M, ASSADI M, El-SHERIF N (1995) Electrophysiologic effects of cocaine on subendocardial Purkinje fibers surviving 1 day of myocardial infarction. J Cardiovasc Electrophysiol 6:729
14. Nattel S, Carlsson L (2006) Innovative approaches to anti-arrhythmic drug therapy. Nat Rev Drug Discov 5:1034
15. Quan W, Rudy Y (1990) Unidirectional block and reentry of cardiac excitation: a model study. Circ Res 66:367
16. MacLeod K (2014) An essential introduction to cardiac electrophysiology [internet]. Imperial College Press, [cited 2019 Jan 3]
17. Spodick DH, Raju P, Bishop RL, Rifkin RD (1992) Operational definition of normal sinus heart rate. Am J Cardiol 69:1245
18. Fagard R (2002) Athlete's heart. Heart 89(12):1455–1461
19. Jensen PN, Gronroos NN, Chen LY, Folsom AR, Defilippi C, Heckbert SR et al (2014) Incidence of and risk factors for sick sinus syndrome in the general population. J Am Coll Cardiol 64:531
20. Marcum ZA, Amuan ME, Hanlon JT, Aspinall SL, Handler SM, Ruby CM et al (2012) Prevalence of unplanned hospitalizations caused by adverse drug reactions in older veterans. J Am Geriatr Soc 60:34
21. Dakkak W, Rhythm DR (2017) Sick sinus syndrome. StatPearls, Treasure Island
22. Prystowsky EN (1988) The effects of slow channel blockers and beta blockers on atrioventricular nodal conduction. J Clin Pharmacol 28:6
23. Abraham S, Reddy S, Abboud J, Jonnalagadda K, Ghanta SK, Kondamudi V (2010) Brief, recurrent, and spontaneous episodes of loss of consciousness in a healthy young male. Int Med Case Rep J 3:71–76
24. Barold SS, Hayes DL (2001) Second-degree atrioventricular block: a reappraisal. Mayo Clin Proc 76:44
25. Jones WM, Rhythm NL (2018) Atrioventricular block, second-degree. StatPearls, Treasure Island
26. Adams MG, Pelter MM (2003) Ventricular escape rhythms. Am J Crit Care 12(5):477–478
27. SCHER AM, RODRIGUEZ MI, LIIKANE J, YOUNG AC (1959) The mechanism of atrioventricular conduction. Circ Res 7:54
28. Narula OS, Scherlag BJ, Samet P, Javier RP (1971) Atrioventricular block. Localization and classification by his bundle recordings. Am J Med 50(2):146–165
29. Palatini P (1999) Need for a revision of the normal limits of resting heart rate. Hypertension 33:622
30. Whinnett Z, Sohaib SA, Davies DW (2012) Diagnosis and management of supraventricular tachycardia. Br Med J 345:e7769
31. Kannel W, Wolf P, Benjamin E, Levy D (1998) Prevalence, incidence, prognosis, and predisposing conditions for atrial fibrillation: population-based estimates 11Reprints are not available. Am J Cardiol 82:2N
32. Nattel S, Dobrev D (2016) Electrophysiological and molecular mechanisms of paroxysmal atrial fibrillation. Nat Rev Cardiol 13:575
33. Wijffels MCEF, Kirchhof CJHJ, Dorland R, Allessie MA (1995) Atrial fibrillation begets atrial fibrillation: a study in awake chronically instrumented goats. Circulation 92:1954
34. Patel P, Pandya J, Goldberg M (2017) NOACs vs. Warfarin for stroke prevention in Nonvalvular atrial fibrillation. Cureus 9(6):e1395
35. Katritsis DG, Josephson ME (2016) Classification, electrophysiological features and therapy of atrioventricular nodal reentrant tachycardia. Arrhythmia Electrophysiol Rev 5:130
36. Schluter M, Geiger M, Siebels J, Duckeck W, Kuck KH (1991) Catheter ablation using radiofrequency current to cure symptomatic patients with tachyarrhythmias related to an accessory atrioventricular pathway. Circulation 84:1644
37. Keating L, Morris FP (2003) Electrocardiographic features of Wolff-Parkinson-white syndrome. Emerg Med J 20:491
38. Sohinki D, Obel OA (2014) Current trends in supraventricular tachycardia management. Ochsner J 14(4):586–595
39. Benito B, Guasch E, Rivard L, Nattel S (2010) Clinical and mechanistic issues in early repolarization: of normal variants and lethal arrhythmia syndromes. J Am Coll Cardiol 56:1177

8

40. de Luna AB, Coumel P, Leclercq JF (1989) Ambulatory sudden cardiac death: mechanisms of production of fatal arrhythmia on the basis of data from 157 cases. Am Heart J 117(1):151–159

41. Echt DS, Liebson PR, Mitchell LB, Peters RW, Obias-Manno D, Barker AH et al (1991) Mortality and morbidity in patients receiving encainide, flecainide, or placebo. The cardiac arrhythmia suppression trial. N Engl J Med 324:781

42. Al-Khatib SM, Stevenson WG, Ackerman MJ, Bryant WJ, Callans DJ, Curtis AB et al (2018) AHA/ACC/HRS guideline for management of patients with ventricular arrhythmias and the prevention of sudden cardiac death. J Am Coll Cardiol 15(10):e190–e252

43. Muser D, Santangeli P, Castro SA, Pathak RK, Liang JJ, Hayashi T et al (2016) Long-term outcome after catheter ablation of ventricular tachycardia in patients with nonischemic dilated cardiomyopathy. Circ Arrhythmia Electrophysiol 9(10):e004328

44. van Herendael H, Dorian P (2010) Amiodarone for the treatment and prevention of ventricular fibrillation and ventricular tachycardia. Vasc Health Risk Manag 6:465–472

45. Guan Yap Y, John Camm A, Yap YG, Camm AJ (2012) Drug induced QT prolongation and torsades de pointes. Heart 2003;89:1363–1372.

46. Zipes DP, Camm AJ, Borggrefe M, Buxton AE, Chaitman B, Fromer M, Gregoratos G et al (2006) 2006 ACC/AHA/ESC guidelines for management of patients with ventricular arrhythmias and the prevention of sudden cardiac death. J Am Coll Cardiol 114:e385–e484

47. Moe GK, Abildskov JA (1959) Atrial fibrillation as a self-sustaining arrhythmia independent of focal discharge. Am Heart J 58:59

48. Haïssaguerre M, Jaïs P, Shah DC, Takahashi A, Hocini M, Quiniou G, et al (1998) Spontaneous initiation of atrial fibrillation by ectopic beats originating in the pulmonary veins. N Engl J Med 339(10):659–666 [Internet]. Available from: http://www.nejm.org/doi/abs/10.1056/NEJM199809033391003.

The Scientific Rationale of Artificial Pacing

Dominic Gyimah, Ahran D. Arnold, and Zachary I. Whinnett

© Springer Nature Switzerland AG 2019
C. Terracciano, S. Guymer (eds.), *Heart of the Matter*, Learning Materials in Biosciences,
https://doi.org/10.1007/978-3-030-24219-0_9

What You Will Learn in This Chapter

This chapter will comprehensively explore the evidence base and underlying scientific rationale for artificial pacemakers and implantable cardioverter defibrillators (ICDs). This will involve a discussion of the three main categories of artificial pacemakers and a broad overview of the pacing modes available, in addition to a comparison of single vs. dual chamber pacing. Then, the indications of ICDs in both the primary and secondary prevention of sudden cardiac death (SCD) will be examined, before we go on to explore the therapeutic potential of His bundle pacing and leadless pacemakers. In particular, this chapter will critically appraise several large studies, identifying the key factors to consider.

Learning Objectives

- Explain the scientific rationale of single and dual chamber pacing.
- Discuss potential mechanisms of cardiac resynchronisation therapy (CRT).
- Be able to evaluate the use of ICDs in the primary and secondary prevention of SCD.

9.1 Historical Perspective of Cardiac Pacing

Artificial pacemakers, distinct from the heart's endogenous pacemaker – the sino-atrial (sinus) node, are medical devices that facilitate the installation and sustenance of a physiological heart rate and rhythm via the generation of targeted electrical activity. This is most often required due to insufficient performance by the sinoatrial node (causing bradycardia), or a block in the cardiac conduction apparatus. Since the implantation of the first pacemaker in 1958, advances in engineering and computer sciences have catalysed the development of highly compact devices, with low power consumption and enhanced performance. This revolution is best exemplified by the minimised potential for heating and electromagnetic interference in newer devices, meaning patients with implanted pacemakers are now able to undergo MRI imaging [1].

And yet, despite such improvements, critical limitations persist. To name a few, this includes lead failure (requiring replacement), device infection, extraction and secondary tricuspid regurgitation [1]. As such, innovative ideas like the leadless cardiac pacemaker (LCP) have been of great interest as a potential next frontier.

9.2 Pacing Mechanisms

Fundamentally, artificial pacemakers comprise a pulse generator, encompassing the battery and capacitor components, as well as electronics and one or more leads that make contact with the myocardium to deliver a depolarising pulse and detect patterns of cardiac electrical activity [2]. A sensed intracardiac signal may prompt a pacemaker to inhibit output, trigger output, or pace in a different chamber following a timed delay. Pacing occurs when a potential difference is applied between two electrodes, whilst the decision to deliver a pacing impulse is governed by the 'programmed pacing mode' [2].

9.2.1 Categories of Pacemakers

Broadly speaking, there are three main types of pacemakers:

1. *Single Chamber System*: a single lead is connected to the generator and implanted into the heart, typically either an atrial or ventricular lead implanted in the right atrium (RA) or right ventricle (RV), respectively. When atrioventricular (AV) conduction is normal, there is, theoretically, no requirement for ventricular pacing, and thus atrial-only ('AAI') pacemakers have been prevalent in the past. Importantly however, with patients usually progressing to requiring two wires over time, AAI systems are rarely used [2]. The ventricular-only ('VVI') pacemaker is useful in patients who have no need for atrial sensing or atrial pacing, which is the case in permanent atrial fibrillation. Therefore, the most common form of single chamber pacemakers is the VVI.

2. *Dual Chamber*: one wire is in the RA and another in the RV, providing sequential contraction of the atria followed by the ventricles [3]. Even if there is sinus node disease alone, where only atrial pacing is initially required, this kind of pacemaker with atrial and ventricular leads is usually implanted [3].

3. *Biventricular Pacing (BVP)*: a third lead, in addition to the right atrial and right ventricular leads, is added to the left ventricle (LV), usually in a branch of the coronary sinus (the venous drainage of the heart). Often termed 'Cardiac Resynchronization Therapy (CRT)', this allows the entire heart to contract more synchronously in patients whose intrinsic contraction is dyssynchronous [2, 3].

9.2.1.1 Pacing Modes

The pacing mode of the device is described using a four or five letter code (as mentioned earlier, e.g., AAI, VVI, etc.) as follows:

- *First Letter*: identifies the chamber *paced*.
- *Second Letter*: identifies the chamber *sensed*.
- *Third Letter*: describes the device *response* to sensed events (I for inhibit, T for trigger, D for both).
- *Fourth Letter*: indicates whether the *rate response* is on.
- *Fifth Letter* (where appropriate): identifies whether *multisite* pacing is employed.

A summary of different pacing modes is shown in ◘ Table 9.1:

9.2.2 Pacing Indications

9.2.2.1 Bradyarrhythmia Indications

Bradyarrhythmias represent a major indication for the application of pacing and often arise from intrinsic insufficiency of the sinus node itself, such as in sinoatrial block or sinus arrest. Notably, more than half of all pacemakers are implanted for this fundamental issue of sinus node dysfunction (SND) [4]. In these patients, dual-chamber pacing is often employed for two key reasons:

- VVI pacing is not used, as it is associated with so-called 'pacemaker syndrome', leading to 1:1 ventriculoatrial conduction or loss of AV synchrony, which can result in AF.
- AAI pacing is typically not used, as the risk for AV nodal disease in patients with SND is reported to be 0.6–2% annually [4].

☐ Table 9.1 A summary of different pacing modes employed in artificial cardiac pacemakers and their common indications and functions

Pacing mode	Indication	Function
DDD	Intact sinus function but impaired atrioventricular (AV) conduction	Sinus activity is sensed and will trigger ventricular pacing after a programmed AV delay
DDDR	Rate response is added when both sinus and AV nodal functions are impaired	The rate response feature provides a chronotropic pacing intervention
VVI and VVIR	Ventricular-only pacing used in patients with chronic atrial fibrillation or bradycardias	VVI(R) hafts the relay of atrial arrhythmias into the ventricles
AAIR	Atrial demand pacing is reserved for isolated sinus node dysfunction with normal AV nodal conduction	Used to prevent the dissemination of aberrant cardiac rhythms to the ventricular myocardium
VOO/DOO	Asynchronous modes with no sensing or rate modification. Used temporarily to prevent oversensing	Prevent sensing of extrinsic electrical activity that may be "misinterpreted' as native cardiac events, e.g., surgery

Information from [2]

Alternatively, bradyarrhythmias may originate from AV nodal block. Here, pacing is indicated for either third-degree blocks – where no beats are conducted, there is complete AV dissociation and a high risk of sudden cardiac death – and second-degree type II blocks – where some beats are not conducted to the ventricles, without PR elongation in the preceding PR intervals [5, 6]. Pacing here can provide prognostic benefit, lowering the risk of SCD. Other forms of heart block (first degree and Mobitz type I second degree) may also be considered for pacing where they are heavily symptomatic. Dual-chamber pacing is utilized here to maintain atrioventricular synchrony, even when the sinus node functions normally – as the atria must be sensed.

9.2.2.2 Pacing Therapy for Heart Failure

In patients with left bundle branch block, the lateral wall of the left ventricle is activated late, which translates to a delayed contraction of this wall. When regions of the left ventricle contract with delay, they fail to contribute to stroke volume and thus cardiac output [7]. This is known as ventricular dyssynchrony, and results in prestretch of the LV free wall, delayed contraction of the LV free wall, inefficient contraction, and potentially end-systolic mitral regurgitation [7]. In patients with left ventricular impairment, these effects result in worsening of heart failure symptoms and higher mortality.

To combat this dyssynchrony, biventricular pacing improves left ventricular function by pacing to restore electrical coordination and mechanical efficiency. This is achieved by the strategic placement of an epicardial lead in a branch of the coronary sinus resulting in a pacemaker that is capable of simultaneously pacing the lateral left ventricular wall and the right ventricular apex to reinstate interventricular and intra-ventricular synchrony [7].

9.2.3 Are Atrial Leads Necessary?

In patients with atrioventricular block who require ventricular pacing, there is a question over whether ventricular-only pacing with a single chamber VVI pacemaker is adequate or whether additional benefit is provided by the addition of an atrial lead to maintain atrioventricular synchrony. Data from UKPACE, and a subsequent meta-analysis of similar trials, did not find a significant difference in mortality between these two strategies. However there is observational evidence that atrial fibrillation and heart failure can be prevented with an atrial lead. Given the use of an atrial lead in preventing the symptoms of pacemaker syndrome, a dual chamber pacemaker with atrial and ventricular leads are typically preferred [5, 8–10].

9.2.4 Downsides of RV Pacing

The dual chamber and VVI implantable defibrillator (DAVID) trial established that in patients indicated for implantable cardioverter defibrillator (ICD) therapy with a left ventricular ejection fraction ≤40%, right ventricular pacing was in fact clinically detrimental, leading to an increase in the primary endpoints of death or hospitalisation for HF compared with pacemaker programming that avoids ventricular pacing: 'back-up' only pacing [11]. One potential reason for this finding could be that unlike the endogenous His-Purkinje system, RV pacing activates the interventricular septum before the LV lateral wall, promoting a left bundle branch block (LBBB) morphology on ECG via the propagation of electrical wavefronts from the sternum. LV dyssynchrony ensues, with mismatched timing between the chamber walls serving to lessen overall left ventricular function [12]. In response to this, algorithms that avoid or minimise RV pacing have been developed for dual-chamber pacemakers [12]. However in some patients ventricular pacing is necessary, so biventricular pacing has been proposed for these patients to mitigate the dyssynchrony caused by ventricular pacing.

9.2.5 Evidence Base for CRT

Much of the evidence base for the use of biventricular pacing arises from the CARE-HF study, which randomised patients to optimal medical therapy with or without CRT-P (CRT with a pacemaker), demonstrating that CRT-P significantly reduced all-cause mortality and unplanned hospitalisations for major cardiovascular events [9]. Around the same time, the COMPANION trial also established CRT as a beneficial treatment in heart failure, particularly those with a New York Heart Association (NYHA) functional class of III or IV, impaired LV function and a wide QRS complex [13].

Since these trials further evidence has been generated from the RAFT and MADIT-CRT studies, among others, demonstrating an improvement in symptoms and mortality with biventricular pacing compared to optimal medical therapy. In patients with a high burden of right ventricular pacing and mild ventricular impairment, ventricular dyssynchrony is problematic in a similar way to left bundle branch block, since both result in slow activation of the left ventricle and, thus, late activation of the lateral ventricular wall. Biventricular pacing for this indication was studied in the BLOCK-HF trial, which found a 25% reduction in a composite of multiple measures of heart failure deterioration including death, and enlargement of the left ventricle. However, this reduction was found to be mainly due to changes in echocardiographic parameters (as opposed to clinical outcomes). This trial has resulted in guideline recommendations, in the US, for biventricular pacing in this population, however [14].

9.3   What We Don't Know: Optimising BVP to Clinical Need

It is noteworthy that the clinical response to BVP is complex, with several different end-points used in key trials, including NYHA functional class, echocardiographic changes, ECG changes and hospitalisation rates. In this context, optimising BVP to maximise its clinical efficacy remains a challenge, particularly in certain demographic cohorts. In an individual patient meta-analysis of five randomised control trials comparing BVP with either no active device or a defibrillator, it was found that only QRS duration accurately predicted the magnitude of BVP on heart failure hospitalisation and mortality [15]. The 'cut-off' for response to CRT in this meta-analysis was between 130 and 150 ms QRS duration, below which BVP may even have worsened outcomes.

Non-LBBB QRS morphology on the other hand, failed to predict clinical response [15]. Therefore, it remains unclear whether BVP should be extended to non-left bundle branch block (LBBB) patients or those with a narrow QRS complex.

9.4   Study Design Issues in CRT Trials

For a class I CRT indication, current guidelines require the presence of LBBB if the QRS complex is broad (>120 ms), with CRT contraindicated when the QRS is <120 ms. Therefore electrical dyssynchrony, manifested as QRS prolongation, is the current target for patient selection for BVP. Mechanical dyssynchrony, measured using echocardiography, was previously proposed as an alternative method for identifying patients with a narrow QRS complex that could benefit from BVP. This recommendation was even incorporated into clinical guidelines. The circumstances leading up to this provide a valuable lesson in study design [15, 16].

Concordantly, the parameters of atrioventricular (AV) and interventricular (VV) delay become increasingly important [4].

Initial studies demonstrating the beneficial effects of BVP in patients with dyssynchronous contraction but narrow QRS complexes suffered from flawed methodology, usually due to either no control arm or no randomisation [17]. Although control arms were introduced in subsequent studies, design flawes persisted. These trials can broadly be stratified into three groups:

- Non-randomised and non-blinded
- Randomised and non-blinded
- Randomised and blinded

The results of these studies were then compared by methodology, with variation in randomisation and blinding associated with contrasting efficacy conclusions [17]. As shown in ◘ Fig. 9.1, the non-randomised and non-blinded trials elucidated a consensus positive mean difference, suggesting improvement from BVP.

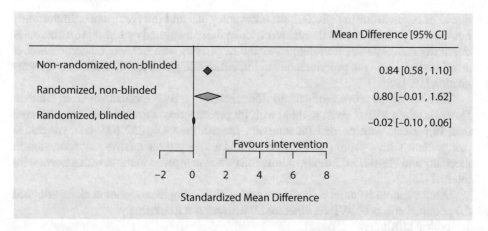

◻ Fig. 9.1 Meta-analysis of CRT intervention studies stratified by the presence of bias-resistance factors (randomisation status and blinding status). Key: red = non-randomised, non-blinded; yellow = randomised, non-blinded; green = randomised, blinded. A mean difference ≥2 favours intervention. (Diagram from [17])

This is explained by patients (with short QRS and mechanical dyssynchrony) being aware that they were reeiving the BVP intervention before being asked they felt better, introducing bias [17]. Interestingly, as soon as patients were randomised and blinded, this phenomenon was diminished, with no siginificant change in symptoms observed with BVP in this group of patients (narrow QRS and mechanical dyssyncrhony). This question was subsequently addressed in a blinded RCT: the EchoCRT trial. This trial recruited patients with evidence of mechanical dyssychrony but no evidence of electrical dyssynchrony (i.e. patients with a narrow QRS complex). This study demonstrated a worse prognosis with BVP compared to medical therapy, including higher mortality.

9.5 BVP: Potential Mechanisms of Benefit

Observed beneficial effects from BVP in heart failure patients include [12]:
- Improvements in LV systolic function and pressure generation
- Increased systolic blood pressure
- Increased stroke volume (and thus reduced LV end-systolic volume)
- Improved cardiac work

Whilst the exact mechanosignalling effects of BVP on the myocardium are yet to be elucidated, a body of evidence supports the following phenomena:
- BVP improves normal and β-adrenergic cardiomyocyte function and Ca^{2+} handling.
- BVP reduces the prolongation in action potential duration observed in the lateral ventricular wall in diastolic heart failure.
- BVP mediates the synchronisation of interventricular activity and apical-to-basal right ventricular activation [18].

A combination of the aforementioned factors causes a leftward shift of LV pressure-volume loops [19]. As previously discussed, BVP is indicated in patients with LBBB chiefly due to its resynchronising effects, both intraventricular and interventricular. Importantly however, BVP also shortens the atrioventricular delay instigated by LBBB [18]. This delay classically manifests as PR prolongation and in patients with ischaemic heart disease, is associated with a worse prognosis due to impairing filling and pre-diastolic mitral regurgitation [18, 19].

Thus far, it has proven difficult to differentiate one BVP mechanism from another. However, when BVP is given to those with PR prolongation, a greater benefit is observed than in patients with normal PR intervals. Indeed, the COMPANION trial concluded that patients with a prolonged PR interval had a 17% greater relative risk reduction in mortality and HF-related hospital admissions when compared to those with a normal PR interval [13].

Using a haemodynamic computer model of the failing heart, Jones et al. investigated the potential effects of BVP on different LV activation states, namely:

1. Typical LBBB.
2. LV activation time of 110 ms (CRT-110), which represents the typical electrical resynchronisation profile obtained with current CRT.

These simulations produced two key findings. Firstly, shortening AV delay is an important mechanism in BVP and in fact may be the dominant mechanism. Secondly, there is scope for further ventricular resynchronization, beyond that provided by conventional BVP, to further improve haemodynamic benefits: an extra 50 ms reduction in LV activation time would deliver an overall effect size equivalent to 53% of the total improvement observed with current methods of delivering CRT [20] (◘ Fig. 9.2).

This model allowed the independent quantification of any potential effects AV delay optimisation (shortening) and ventricular resynchronisation may exert, with both showing beneficial effects on acute haemodynamic function. Moreover, the simulations verified

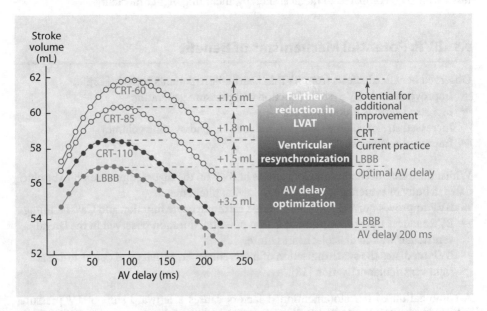

◘ **Fig. 9.2** The relative contributions of AV delay optimisation and ventricular resynchronisation to the improvements in cardiac functional capacity exerted by CRT. (Image from [20])

that the relative importance of both mechanisms is dependent on the baseline PR interval and magnitude of reduction in LV activation time with CRT. While these factors are patient-specific, the results of this study suggest that improving LV preload through the minimisation of AV delay is the foremost mechanism by which CRT ameliorates cardiac function when delivered to patients in sinus rhythm [20].

9.6 Implantable Cardioverter Defibrillators (ICDs)

Implantable cardioverter defibrillators (ICDs) are devices inserted into the body that are capable of 'cardioverting' and 'defibrillating', meaning the targeted delivery of electrical current to the myocardium to terminate life-threatening arrhythmias [21, 22] to prevent sudden cardiac death.

Sudden cardiac death (SCD) refers to unexpected death from a cardiac cause, occurring 1 hour within the onset of symptoms [23]. SCD can be due to mechanical events such as acute failure of heart pumping, myocardial infarction or cardiac tamponade. However, the most common causes of SCD are arrhythmias, and such deaths can be referred to as sudden arrhythmic deaths. Two arrhythmias are typically responsible for sudden arrhythmic death: ventricular tachycardia (VT) and ventricular fibrillation (VF), which together can be referred to as malignant ventricular arrhythmias. VF rarely spontaneously terminates and thus inevitably results in death in the absence of prompt defibrillation either via an ICD or using external pads/paddles connected to an external defibrillator. VT can be fatal itself but can support spontaneous circulation, although often this is a compromised circulation, which results in rapid organ failure if untreated. Furthermore untreated VT can degenerate into VF.

Current guidelines advocate ICD therapy for the 'primary' and 'secondary' prevention of SCD [24]. In the context of ICD therapy, primary prevention refers to ICDs implanted to prevent an initial presentation of SCD, whilst secondary prevention refers to ICDs implanted to recurrence of 'SCD' in those who have already experienced life-threatening VT/VF (when the former episodes of which were successfully treated or terminate spontaenously: aborted SCD).

9.6.1 ICD Mechanism of Action

ICDs are usually transvenous devices, which means there is an infraclavicularly implanted generator (containing circuity, a battery and a capacitor) connected to a defibrillator lead, which is a wire passed through the venous system into the right ventricular cavity, with an electromagnetic coil in contact with the right ventricular myocardium. The lead is passed to the right ventricle by puncturing the subclavian vein, or one its branches, and passing the lead through the supervior vena cava and right atrium before reaching the right ventricle. Defibrillation occurs by passing a current across the heart between the coil and the generator. Such transvenous ICDs are capable of performing bradyarrhythmia pacing and anti-tachycardia pacing in addition to defibrillation. Sub-cutaneous defibrtillators are a more recent development where the generator is placed under the axilla and the lead is placed above the sternum (under the skin) provide coil-generator vector that encompassses the heart, without the need for leads within the venous system or the endocardial cavities of the heart itself. Transvenous ICDs can be combined with cardiac

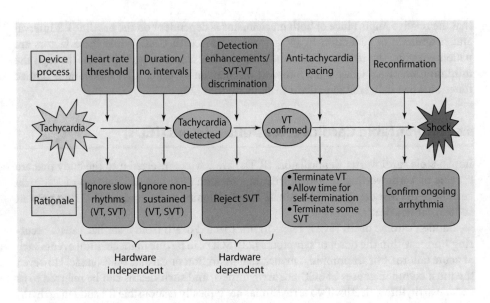

Fig. 9.3 Diagram showing the sequence of events involved in the treatment of ventricular arrhythmias by an ICD. (Diagram from [25])

resynchronisation therapy to form 'CRT-D' devices with defibrillator lead in the right ventricle and pacing lead in a coronary sinys branch. The treatment of ventricular arrhythmia by an ICD involves a coordinated sequence of events (☐ Fig. 9.3). First the rapid heart rate of VT/VF exceeds the programmed minimal heart rate threshold that helps the ICD to distinguish VT/VF from sinus tachycardia or other slower arrhythmias. This is referred to as 'sensing'. A period of time then elapses before ICD therapy is administerd to ensure that VT of short duration (non-sustained VT) is not unnecessarily treated. Then a series of algorithms are applied to the sensed signals to try to distinguish life-threatening, ICD-treatment-requiring VT/VF from AF/SVT where ICD treatments would not be appropriate. Finally treatment is adminstered in the form of ICD 'therapies'. There are 2 kinds of ICD therapy: anti-tachycardia pacing (ATP), which is where the myocardium is stimulated rapidly, or defibrillatory 'shocks' where a high-energy current is passed through the heart. The typical programmed sequence is for attempts of ATP for a short duration with the last resort being shocks. Finally the resulting rhythm is checked for success of the therapy [25].

When ICDs therapies (shocks and/or ATP) are delivered in response to correctly sensed VT or VF, they are called 'appropriate' shocks as this represents the ICD functioning as intended, potentially rescuing patients from death. However, ICDs often deliver therapies in the absence of ventricular arrhythmia. Such therapies are 'inappropriate'. For example, if a patient with an ICD exercises sufficiently vigorously that their heart rate falls into the detection zone for a long period, with a failure of discrimination algorithm to realise it is sinus tachycardia, then a shock will be delivered that is not treating a life-threatening arrhythmia but instead is causing distress to a patient who was otherwise feeling fine. There are several other phenonemena that can result in inappropriate shocks. Any cause of a rapid heart rate, including atrial fibrillation, supraventricular tachycardia (SVT) or sinus tachycardia, can be mistaken for VT/VF by an ICD if discrimination algorithms are unable to distinguish them. Any signal that resembles a fast heart rate will similarly be identified as VT/VF. This can occur in several ways. T waves may be sensed as R waves alongside actual R waves, resulting in doubling of the sensed rate. Muscle activity in the shoulder (myopotentials), defibrillator lead integrity (fracture) or electromagnetic interference (EMI) can produce signals that resemble VT/VF. Thus inappropriate shocks occur in 8–40% of patients with ICDs [22, 25].

Inappropriate shocks are highly undesirable since they cause patients pain and distress and even serious psychological sequelae including post-traumatic stress disorder, but crucially without terminating an otherwise fatal arrhythmia (as there is no fatal arrhythmia present in the first place). Apart from the psychological harm, recent studies suggest that inappropriate shocks cause cardiovascular harm and death. In the MADIT-RIT study, adjusting ICD detection algorithms such that therapy was restricted so that therapy was only delivered at high heart rates of >200 bpm was associated with a twofold reduction of inappropriate shocks and a 55% reduction in all-cause mortality [26]. Elimination of inappropriate shocks remains a key goal in ICD development.

9.6.2 ICDs in Sudden Cardiac Death Prevention

9.6.2.1 Secondary Prevention

Patients who have already survived an aborted SCD event, either due to successful treatment or spontaneous termination of an otherwise fatal arrhythmia, are at high risk of further life-threatening arrhythmias [27]. Pharmacotherapy using anti-arrhythmic drugs such as amiodarone have been used to prevent recurrence of ventricular arrhythmias, with limited success. AVID (Anti-arrhythmics versus implantable defibrillators) was the first major comparative study of drug and ICD therapy, enrolling patients previously resuscitated for VF or cardioverted for sustained VT [28]. These patients were randomised to ICD therapy or class III anti-arrhythmics (primarily amiodarone), with a significant mortality reduction observed in the ICD patients when compared to the anti-arrhythmic group in the three-year follow-up [28].

This conclusion was then reinforced by meta-analysis of the early key studies, of which AVID was one (AVID, CASH and CIDS), concluding that a significant reduction in all-cause and arrhythmic mortality results from the usage of ICD therapy in the secondary prevention population [29], resulting in a storng recommendation for the use of secondary prevention ICDs in international guidelines. The recommendations for secondary prevention ICD use span multiple conditions that predispose towards arrhythmic death, including myocardial infarction, infiltrative disorders, cardiomyopathies, idiopathic syndromes and channelopathies.

9.6.2.2 Primary Prevention

As previously mentioned, the primary prevention ICD population refers to patients who are suspected to be at high-risk of arrhythmic death from VT/VF but have not yet suffered a sustained ventricular arrhythmia. The predisposing conditions are the same as aforementioned for secondary prevention population (myocardial infarction, channelopathies etc.) but the criteria for offering an ICD are different. In the secondary prevention population, the ICD indication is clear as the risk of arrhythmic death (that could be prevented by an ICD) is high for all individuals who have suffered aborted SCD, but with the primary prevention population criteria must be applied. All patients with predisposing conditions cannot be offered ICDs for several reasons. Firstly, the risk of arrhythmic death is not evenly distributed among the patients with arrhythmia-inducing conditions: many patients will never suffer an arrhythmia and thus ICD implantation can only offer them harm. Secondly, it is not economically feasible as the primary prevention population would be too large. Thus high risk features are sought to identify a particularly at-risk subpopulation. In the setting of patients with heart failure from either non-ischaemic cardiomyopathy or previous myocardial infarction, the biomarker that is used to identify risk is severe impairment of left ventrcular function as measured using echocardiography. The MUSTT trial randomised patients with prior MI, left ventricular ejection fraction (LVEF) <40%, documented non-sustained VT, and inducible VT to electrophysiology (EP) study-guided management vs. standard medical care [30].

The results suggested a survival benefit only in patients within the EP study-guided therapy arm who received an ICD, with no similar benefit observed in the anti-arrhythmic cohort.

The randomised MADIT study recruited patients with prior MI, LVEF <30%, documented non-sustained VT and inducible VT not suppressible with procainamide to either ICD or medical therapy. The subsequent 54% relative reduction in all-cause mortality in the MADIT ICD group served to reinforce the findings of MUSTT [31].

Also supporting the findings of MUSTT was the follow-up to MADIT, MADIT-II, which found ICD therapy was associated with a 49% relative reduction in all-cause mortality in a broader population without the need for inducible arrhythmia [31].

However, the largest study to date in the primary prevention of SCD in heart failure patients was the Sudden Cardiac Death in Heart Failure (SCD-HeFT) trial, which randomised 2521 patients with NYHA class II or III HF and LVEF <35% to ICD, amiodarone or placebo therapy. The results showed no significant mortality difference between the amiodarone and placebo groups, with a 23% relative mortality reduction witnessed in the ICD arm [32]. Thus in patients with severe left ventricular impairment there is a strong recommendation in international guidelines to implant an ICD. Outside of heart failure indications, the critiera for channelopathy and cardiomyopathy are much more complex involving frequently-evolving risk scores.

9.7 Where We're Heading

Even though cardiac device therapy is a highly technological field, there are many large remaining challenges that are in the process of being addressed by yet further advances in technology. Below, two important developments are described in the field of cardiac pacing but there are also important advances in the field of defibrillation. Novel ways of preventing inappropraite shocks are in development including the incoporation of haemodynamic signals into device function. The relatively recently invented subcutaneous defibrillators, which, as described above, avoid leads within the ventricular cavities, are already common place.

9.7.1 His Bundle Pacing

Conventional ventricular pacing, described in the early sections of this chapter, involves stimulating the myocardium of the right ventricle (or both the right and left ventricles in biventricular pacing). This produces a non-physiological activaiton pattern. Right ventricular pacing has a similar ECG appearence to left bundle branch block and, as a result, a simularly dyssyncrhonous contraction of the ventricles. This results, overtime, in venticular impairment. His bundle pacing involves direct stimulation of the His-Pukinjie conduction system so that the ventricles are activated identically to native, synchronous contraction.

This method of pacing was first described in 1970 by Narula et al. However, it was not demonstrated experimentally at any significant scale until 2000 [33]. His bundle pacing has been shown to facilitate optimisation of AV delay whilst retaining normal intrinsic ventricular activation, particularly in patients with HF and PR prolongation in the context of a normal QRS or RBBB [33]. Consequently, the His Optimised Pacing Evaluated for Heart Failure (HOPE-HF) trial currently underway will be critical in further evaluating the clinical outcomes from His bundle pacing. Importantly though, this technique is now without limitations:

1. Success rates for His lead implantation have been as low as 60% without dedicated tools.
2. Pacing thresholds can be relatively high compared to current thresholds, suggesting elevated energy demands. The implications on battery longevity are ultimately yet to be elucidated.

His bundle pacing can even reverse left bundle branch block as a potentially superior alternative to cardiac resynchronisation therapy [33].

9.7.2 Leadless Pacemakers

Despite improvements in contemporary pacing devices, short-term complication rates remain as high as 9.5–12.4%, running the gamut from pneumothorax and cardiac tamponade to pocket haematoma and lead dislodgement [34]. Moreover, around 0.7 to 2.4% of patients suffer from serious complications related to the subcutaneously-placed pulse generator, including skin erosion, pocket infection and septicaemia [35]. In response to this, leadless pacemakers were conceptualised in the 1970s to address some of these issues; however, their use was ultimately hindered by advancements in battery technologies to increase device longevity, miniaturization of device components, mechanisms to minimise power consumption, and catheter-based delivery tools to enable safe administration to the myocardial wall [34, 35].

The first human trial, LEADLESS, had an overall complication-free rate of 94% at 90 days follow-up amongst the 33 subjects enrolled [36]. The successor study, LEADLESS II, then involved 527 patients in order to assess safety and efficacy with a higher-degree of reliability. This time, the device was implanted successfully in 95.85% of patients, with device-related adverse events arising in only 34 patients [37].

However, current leadless pacemaker systems are RV single-chamber pacemakers only. These are indicated in only a minority (15–30%) of total pacemaker recipients in the developed world, namely those with chronic AF and AV block [34]. Nonetheless, if leadless pacing systems prove durable and as effective as transvenous pacing, its availability may broaden to include patients who otherwise might receive dual-chamber devices.

> **Take-Home Message**
>
> - Bradyarrhythmias can be treated effectively with permanent pacemakers that can sense and stimulate cardiac activity.
> - Biventricular pacing is effective in improving mortality and morbidity in patients with heart failure and broad left bundle branch block.
> - ICDs are indicated in both the primary and secondary prevention of sudden arrhythmic death.

References

1. Austin C, Kusumoto F (2016) Innovative pacing: recent advances, emerging technologies, and future directions in cardiac pacing. Trends Cardiovasc Med 26:452
2. Mulpuru SK, Madhavan M, McLeod CJ, Cha Y-M, Friedman PA (2017) Cardiac pacemakers: function, troubleshooting, and management. J Am Coll Cardiol 69:189
3. Kotsakou M, Kioumis I, Lazaridis G, Pitsiou G, Lampaki S, Papaiwannou A et al (2015) Pacemaker insertion. Ann Transl Med 3(3):42
4. Das MK, Dandamudi G, Steiner HA (2009) Modern pacemakers: hope or hype? PACE. Pacing Clin Electrophysiol 32:1207
5. Tse HF, Lau CP (2006) Clinical trials for cardiac pacing in bradycardia: the end or the beginning? Circulation 114:3
6. Honarbakhsh S, Hunter L, Chow A, Hunter RJ (2018) Bradyarrhythmias and pacemakers. BMJ 360:k642
7. Madhavan M, Mulpuru SK, McLeod CJ, Cha Y-M, Friedman PA (2017) Advances and future directions in cardiac pacemakers. J Am Coll Cardiol 69:211
8. Thibault B, Ducharme A, Harel F, White M, Omeara E, Guertin MC et al (2011) Left ventricular versus simultaneous biventricular pacing in patients with heart failure and a qrs complex $\geq$120 milliseconds. Circulation 124:2874
9. Cleland JGF, Daubert J-C, Erdmann E, Freemantle N, Gras D, Kappenberger L et al (2005) The effect of cardiac resynchronization on morbidity and mortality in heart failure. N Engl J Med 352:1539
10. Toff WD, Skehan JD, De Bono DP, Camm AJ (1997) The United Kingdom pacing and cardiovascular events (UKPACE) trial. Heart 78(3):221–223
11. Wilkoff BL, Cook JR, Epstein AE, Greene L, Hallstrom AP, Hsia H et al (2002) Dual-chamber pacing-or ventricular backup pacing in patients with an implantable defibrillator: the Dual Chamber and VVI Implantable Defibrillator (DAVID) trial. J Am Med Assoc 288(24):3115–3123
12. Leyva F, Nisam S, Auricchio A (2014) 20 years of cardiac resynchronization therapy. J Am Coll Cardiol 64:1047
13. Bristow MR, Saxon LA, Boehmer J, Krueger S, Kass DA, De Marco T et al (2004) Cardiac-resynchronization therapy with or without an implantable defibrillator in advanced chronic heart failure. N Engl J Med 350:2140
14. Curtis AB, Worley SJ, Adamson PB, Chung ES, Niazi I, Sherfesee L et al (2013) Biventricular pacing for atrioventricular block and systolic dysfunction. N Engl J Med 368:1585
15. Steffel J, Robertson M, Singh JP, Abraham WT, Bax JJ, Borer JS et al (2015) The effect of QRS duration on cardiac resynchronization therapy in patients with a narrow QRS complex: a subgroup analysis of the EchoCRT trial. Eur Heart J 36:1983
16. Pacing C (2013) 2013 ESC guidelines on cardiac pacing and cardiac resynchronization therapy. Eur Heart J 15(8):1070–1118
17. Jabbour R, Shun-Shin M, Finegold J, Sohaib A, Cook C, Whinnett Z et al (2014) 42 meta-analysis identifying the source of conflict between different trial reports on the effect of CRT in heart failure with narrow QRS complexes. Heart 100:A23.1
18. Houthuizen P, Bracke FALE, Van Gelder BM (2011) Atrioventricular and interventricular delay optimization in cardiac resynchronization therapy: physiological principles and overview of available methods. Heart Fail Rev 16:263

9

19. Cho H, Barth AS, Tomaselli GF (2012) Basic science of cardiac resynchronization therapy: molecular and electrophysiological mechanisms. Circ Arrhythm Electrophysiol 5:594

20. Jones S, Lumens J, Sohaib SMA, Finegold JA, Kanagaratnam P, Tanner M et al (2017) Cardiac resynchronization therapy: mechanisms of action and scope for further improvement in cardiac function. Europace 19(7):1178–1186

21. Dosdall DJ, Huang J, Ideker RE (2009) Mechanisms of defibrillation. In: Ventricular arrhythmias and sudden cardiac death. Hoboken, New Jersey, USA: Blackwell Publishing

22. Miller JD, Yousuf O, Berger RD (2015) The implantable cardioverter-defibrillator: an update. Trends Cardiovasc Med 25:606

23. Srinivasan NT, Schilling RJ (2018) Sudden cardiac death and arrhythmias. Arrhythmia Electrophysiol Rev 7:111

24. Al-Khatib SM, Stevenson WG, Ackerman MJ, Bryant WJ, Callans DJ, Curtis AB et al (2018) AHA/ACC/HRS guideline for management of patients with ventricular arrhythmias and the prevention of sudden cardiac death. J Am Coll Cardiol 138(13):e210–e271

25. Madhavan M, Friedman PA (2013) Optimal programming of implantable cardiac-defibrillators. Circulation 128:659

26. Moss AJ, Schuger C, Beck CA, Brown MW, Cannom DS, Daubert JP et al (2012) Reduction in inappropriate therapy and mortality through ICD programming. N Engl J Med 367:2275

27. Wathen M (2007) Implantable cardioverter defibrillator shock reduction using new antitachycardia pacing therapies. Am Heart J 153:44

28. Richardson DW, Cobb LA, Pratt CM, Anderson JL, Ehlert F, Epstein AE et al (1999) Causes of death in the Antiarrhythmics Versus Implantable Defibrillators (AVID) trial. J Am Coll Cardiol 34(5):1552–1559

29. Connolly SJ, Hallstrom AP, Cappato R, Schron EB, Kuck K, Zipes DP et al (2000) Meta-analysis of the implantable cardioverter defibrillator secondary prevention trials. AVID, CASH and CIDS studies. Antiarrhythmics vs implantable defibrillator study. Cardiac Arrest Study Hamburg. Canadian Implantable Defibrillator Study. Eur Heart J 21:2071

30. Lee KL, Hafley G, Fisher JD, Gold MR, Prystowsky EN, Talajic M et al (2002) Effect of implantable defibrillators on arrhythmic events and mortality in the multicenter Unsustained tachycardia trial. Circulation 106:233

31. Moss AJ (2003) MADIT-I and MADIT-II. J Cardiovasc Electrophysiol 14:S96

32. Bardy GH, Lee KL, Mark DB, Poole JE, Packer DL, Boineau R et al (2005) Amiodarone or an implantable cardioverter–defibrillator for congestive heart failure. N Engl J Med 352:225

33. Ali N, Keene D, Arnold A, Shun-Shin M, Whinnett ZI, Sohaib SA (2018) His bundle pacing: a new frontier in the treatment of heart failure. Arrhythmia Electrophysiol Rev 7:103

34. Merkel M, Grotherr P, Radzewitz A, Schmitt C (2017) Leadless pacing: current state and future direction. Cardiol Ther 6:175

35. Tjong FVY, Reddy VY (2017) Permanent leadless cardiac pacemaker therapy: a comprehensive review. Circulation 135:1458

36. Reddy VY, Knops RE, Sperzel J, Miller MA, Petru J, Simon J et al (2014) Permanent leadless cardiac pacing: results of the LEADLESS trial. Circulation 129:1466

37. Reddy VY, Exner DV, Cantillon DJ, Doshi R, Bunch TJ, Tomassoni GF et al (2015) Percutaneous implantation of an entirely intracardiac leadless pacemaker. N Engl J Med 373:1125

Cardiac Contractility

Fotios G. Pitoulis and Pieter P. de Tombe

© Springer Nature Switzerland AG 2019
C. Terracciano, S. Guymer (eds.), *Heart of the Matter*, Learning Materials in Biosciences,
https://doi.org/10.1007/978-3-030-24219-0_10

What You Will Learn in This Chapter

This chapter will equip you with an understanding of the determinants of cardiac contractility and the changes observed in these within the context of heart failure. We will begin by discussing the fundamentals of cardiac output, the effects of preload and afterload on ventricular function, and their clinical significance.

Learning Objectives

- Understand the main determinants of cardiac performance.
- Appreciate the difference between preload, afterload and contractile state.
- Appreciate the importance of pressure–volume loops.

10.1 What We Know: Cardiac Output

Cardiac output (CO) refers to the volume of blood ejected by the heart in 1 minute, expressed in litres per minute (l/min). CO continuously fluctuates to meet the changing energetic demands of the body and is determined by the product of the stroke volume(SV)(the volume of blood ejected by the heart in each beat) and heart rate (HR), in the following relationship: CO = SV × HR (Eq. 1) [1]. In the first part of this chapter, we will look into the fundamental role of HR in the regulation of CO, and appreciate how SV is governed by preload, afterload and cardiac contractility. The relationship of these parameters in the context of pressure–volume loops will then be discussed.

In response to stress (e.g. exercise, pregnancy, heart failure), a primitive fight-or-flight response is mounted by the adrenergic system, triggering an increase in CO. This is achieved in part through an increase in HR (known as positive chronotropy). Yet, because HR and SV are not independent variables, the effect of HR on CO is not straightforward. In fact, increasing HR decreases SV (◘ Fig. 10.1a). In open-chested dogs, artificially pacing the heart to high HRs causes the CO to increase, reach a plateau and then drop above a threshold value (◘ Fig. 10.1b) [2]. However, the main determinant of CO is venous return and not HR (see *preload*). Generally, maximal CO is expected to occur with the highest possible HR that does not compromise filling [3].

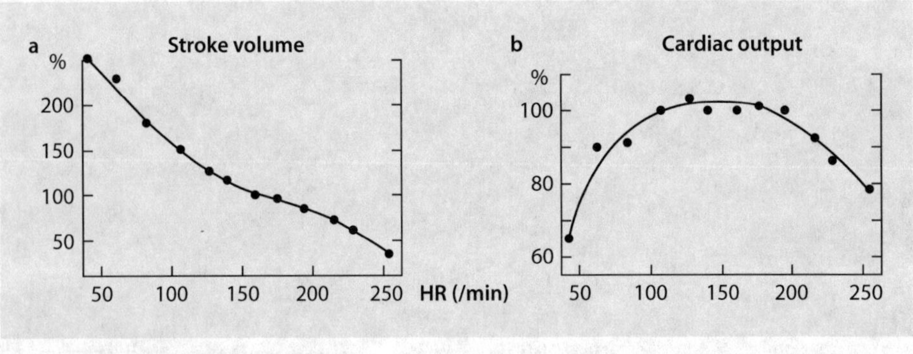

◘ **Fig. 10.1** **a** Relationship between SV and HR, and **b** relationship between CO and HR in open-chest canines. (Adapted from [2])

10.1.1 The Role of Ca^{2+} in Cardiac Contraction

During the cardiac action potential, Ca^{2+} influx via sarcolemmal L-type Ca^{2+}-channels leads to Ca^{2+} release from the sarcoplasmic reticulum and the generation of the whole-cell Ca^{2+} transient (see chapter *excitation–contraction coupling*). The subsequent rise in Ca^{2+} then triggers a molecular interaction between thin (actin) and thick (myosin) myofilaments on sarcomeres, the cardiomyocyte's contractile units [4]. The magnitude of force generated by cardiomyocytes is strongly paired to the intracellular Ca^{2+} concentration, $[Ca^{2+}]_i$ [5].

Specifically, increasing $[Ca^{2+}]_i$ results in a non-linear increase in force, meaning that small changes in $[Ca^{2+}]_i$ can result in functionally significant changes in CO [6]. This steep relationship between $[Ca^{2+}]_i$ and force can be described with a negative logarithmic $[Ca^{2+}]_i$ scale (pCa), characterized by a sigmoid curve organized symmetrically around the half-maximum force at pCa_{50}, as shown in ◻ Fig. 10.2 [6]. Changes in the sarcomere's myofilament Ca^{2+} sensitivity can affect the magnitude of force generated at a given $[Ca^{2+}]_i$. Interventions that affect myofilament Ca^{2+} sensitivity can shift the curve to the left or right, leading to greater or lesser force generation at a given $[Ca^{2+}]_i$, respectively [7, 8]. For example, decreased pH increases pCa_{50} (decreases Ca^{2+} sensitivity), whereas increasing sarcomere length decreases pCa_{50}, thereby increasing Ca^{2+} sensitivity.

10.1.2 Preload

In 1884, Howell and Donaldson demonstrated that increasing the venous return in an isolated mammalian heart–lung preparation increased CO, whilst decreasing it had the opposite effect [7]. A few years later, Otto Frank and Ernest Starling showed that stepwise increases in diastolic volume and pressure increased the magnitude of cardiac contraction [9, 10]. To study these observations *in vitro*, strips of ventricular myocardium are stretched

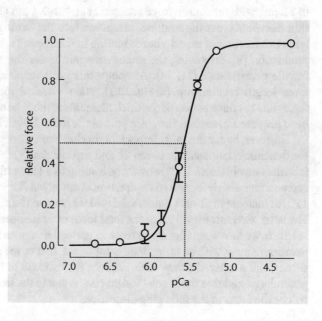

◻ **Fig. 10.2** Force–pCa curve in demembranated ventricular preparations where force has a sigmoidal relationship with $[Ca^{2+}]_i$. Dotted lines correspond to pCa_{50}: the $[Ca^{2+}]_i$ at half-maximal force. pCa_{50} is a measure of Ca^{2+} myofilament sensitivity. Decreasing pCa_{50} corresponds to a rightward shift of the curve and a decrease in Ca^{2+} sensitivity. (Modified from Sun et al. [6])

Fig. 10.3 Muscle length–force relationship in cardiac muscle. Increasing muscle length (and consequently sarcomere length) increases force. Notice steepness of relationship. At 70% of L_{max} force tends towards abscissa. Adapted from [10])

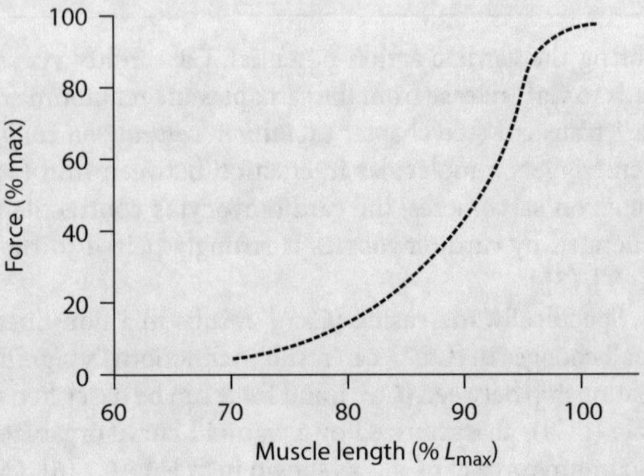

to different lengths and the generated force measured. Application of stretch represents a load on the myocardium, termed *preload*. With electrical stimulation, the myocardium contracts, allowing us to investigate the effects of varying preload on contraction. *Increasing preload increases force generation.* This observation is known as force–length relationship or *Frank–Starling Law* (Fig. 10.3).

At the level of the whole heart, increased preload is represented by greater blood filling of the ventricle, increasing end-diastolic volume (EDV) and thus pressure (EDP). This stretches the myocardium, explaining why increased venous return augments cardiac contraction [8]. So, what is the mechanism behind this phenomenon?

Laser diffraction measurements show that at higher preload, sarcomeres in cardiomyocytes are lengthened [8]. When force is plotted as a function of sarcomere length (SL), increasing the latter results in increase in force [9]. The first mechanism to explain this is known as *filament overlap and interference* [9]. *In vivo*, cardiomyocyte SL ranges from 1.8 to 2.2 μm, with maximum force attained at SL 2.2–2.4 μm [11, 13]. At low SLs (<2.0 μm), thin filaments cross the midline, extending into the 'wrong' half of the sarcomere. This leads to a fraction of myosin heads pulling in the opposite direction, hampering net force production [8]. Stretching the sarcomeres minimizes this interference, increasing contractile performance [11, 14]. Although this partly explains the ascending limb of the force–length relationship (Fig. 10.3), other physical mechanisms such as opposing forces due to compression of the thick filaments at the Z-bands at low muscle lengths have been proposed to explain the steep part of the force–length function [10].

However, by far the most important mechanism underlying the Frank–Starling law is the dynamic relationship between SL and myofilament Ca^{2+} sensitivity [5]. Increasing SL increases myofilament Ca^{2+} sensitivity, leading to a leftward shift in the force–Ca^{2+} curve and an immediately increased contraction at any given $[Ca^{2+}]_i$, shown in Fig. 10.4 [7, 11, 15]. An increase in SL also causes a delayed increase in the amplitude of the Ca^{2+} transient. The latter accounts for ≈30% of the total force increase observed with stretch [12].

Thus, we have seen that an increase in preload increases force generation in a biphasic manner, namely, with an immediate component, that is, the shift in Ca^{2+}myofilament sensitivity, and a delayed component, namely, the increase in Ca^{2+} transient amplitude. It is generally agreed that the Frank–Starling law refers to the immediate component, and this will be the focus of the following discussion.

■ **Fig. 10.4** Force-Ca²⁺ Curve. Increasing Ca²⁺ concentration increases force generation in a non-linear manner. Increasing sarcomere length (i.e. increasing preload) from 1.9–2.04 µm to 2.3–2.5 µm increases the myofilament Ca²⁺ sensitivity leading to greater force at any given Ca²⁺. (Adapted from Hibberd et al. [32])

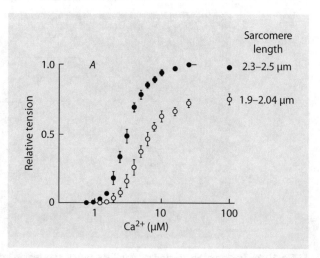

■ **Fig. 10.5**   Interfilament lattice spacing theory of length-dependent activation. For a cross-bridge cycle to occur, myosin heads must bind to myosin binding sites on actin. According to the interfilament lattice spacing theory, when sarcomeres are stretched (from **a** to **b**), the distance between myosin and actin decreases, promoting the cross-bridge cycle and the generation of force

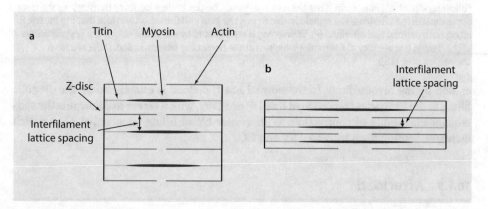

   This leaves us with the task of explaining the increased myofilament Ca²⁺ sensitivity at increased SLs. One hypothesis is that increasing SL decreases interfilament lattice spacing. As the volume of a cardiomyocyte is constant, increases in length are accompanied by a decrease in the spacing between subcellular components, such as the distance between filaments in the sarcomeres. This enhances the probability of an interaction between thick and thin filaments, thereby increasing force generation at a given $[Ca^{2+}]_i$ as illustrated in ■ Fig. 10.5 [13].

   The interfilament lattice space theory may be insufficient to explain the length-dependent increase in myofilament sensitivity on its own. X-ray diffraction patterns suggest that increasing SL results in a superior orientation of myosin heads on the thick filament backbone, which may contribute to the increase in force [14].

   We have therefore seen that the muscle length–force relationship is a consequence of 'physical' and 'activating' factors [10]. The former refers to changes in myofilament overlap and opposing forces with muscle length changes, whereas the latter refers to changes in myofilament sensitivity to Ca²⁺. Overall, increased preload leads to increased force gen-

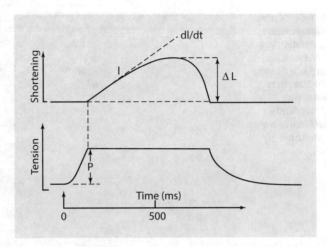

■ **Fig. 10.6** Tracing of an isolated cat papillary muscle shortening in the presence of afterload. Following electrical stimulation (time point 0), the muscle begins to develop force (tension). As the force approximates the afterload, the muscle is able to pull the load and thus shorten. Note that the muscle is lifting both preload and afterload (*P*). Shortening is represented by Δ*L*, whilst *dl/dt* is the first derivative of shortening (i.e. velocity of shortening), highest at the beginning of contraction. (Adapted from Sonneblick et al. [15])

eration by the myocardium. In the normal heart, preload is established by the diastolic filling of the ventricles (represented by EDP or EDV), which serves to determine the subsequent magnitude of contraction. With greater blood filling, myocardial fibre stretch increases, leading to an increased SV and CO.

10

10.1.3   Afterload

Another determinant of cardiac performance is *afterload*, defined as the mechanical load imposed on the ventricle during the ejection phase. At the level of the whole heart, afterload is broadly represented by the aortic pressure, that is, the pressure the ventricles must produce to eject blood into the aorta [1]. Once ventricular pressure exceeds aortic pressure, the aortic valve opens to facilitate the propulsion of blood into the circulation [9]. At the cardiomyocyte level, once the developed force starts approximating the afterload, sarcomeres begin to shorten and contraction shifts from isometric (constant length) to isotonic (changing length), as shown in ■ Fig. 10.6 [15].

As we previously described, preload can be applied on isolated ventricular strips by stretching the tissue, typically done by hanging weights on the preparation. In more complex experimental setups (utilizing a stop lever), weights can be added such that they will only be encountered once the muscle starts contracting, that is, representing the afterload. Thus, before the beginning of contraction, the muscle is stretched by the preload; however, after the onset of contraction, the muscle will lift both the preload and afterload, with total lifted load *P* = *preload* + *afterload* [15]. If cardiac performance is defined as the mechanical work performed by the muscle, then:

$$\text{Work} = P \times \Delta L$$

where Δ*L* is the total shortening of the myocardium.

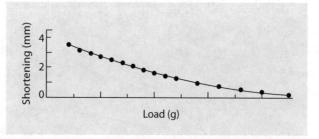

Fig. 10.7 Effect of afterload on shortening of isolated cat papillary muscles. (Adapted from Sonneblick et al. [15])

By varying preload and afterload, their influence on cardiac performance can be deciphered. The first observation is that for a given preload, increasing afterload diminishes ΔL in a monotonic function due to the muscle contracting against greater resistance. Eventually, the muscle is unable to pull the load (i.e. it cannot shorten; isometric contraction), as shown in **Fig. 10.7** [19–21]. Therefore, when afterload is so high that the muscle is contracting isometrically (i.e. $\Delta L = 0$), $W = 0$. Secondly, if we allow afterload to approach 0, then ΔL will be maximal, as the muscle is shortening against ≈ 0 resistance; work will also be 0 as $P \approx 0$.

Understanding the role of afterload is important clinically as pharmacological agents and/or mechanical devices may be used to alter it, increasing the myocardial contractile capacity (and thereby CO), whilst decreasing the workload (and by extension energetic demands) on the heart. In general, the more loaded the tissue, the higher the metabolic demands so unloading the heart (e.g. with assist devices) can limit energy and O_2 consumption [16], which is important both acutely (e.g. acute myocardial infarction) and chronically (e.g. heart failure). A completely unloaded left ventricle has an average consumption of 2 ml O_2 per minute per 100 g of muscle [17].

10.1.4 Cardiac Contractility

'*Contractility*' refers to the inotropic capacity of the myocardium and describes the ability of the heart to eject blood at a given preload and afterload [22, 23]. If CO is plotted as a function of left-ventricular end-diastolic pressure (LVEDP), the graph that is obtained is called Frank–Starling or ventricular function curve and shows a curvilinear CO increase with increasing LVEDP [8]. These curves show that for any given LVEDP, an increase in myocardial contractility shifts the curve upwards and to the left (**Fig. 10.8**).

The principal determinant of the cardiac inotropic state (i.e. contractility) is the autonomic nervous system (ANS). Specifically, the sympathetic arm of the ANS mounts the cardiac adrenergic response via the release of cardiac-acting catecholamines including adrenaline and noradrenaline. The former is synthesized in the adrenal medulla and secreted into the circulation, whilst the latter is released directly into the cardiac interstitium by sympathetic fibres [19]. At the cellular level, adrenergic stimulation regulates the excitation–contraction coupling machinery in the following ways:
1. Increased Ca^{2+} transient amplitude, meaning greater activation of cross-bridges and stronger contraction (positive inotropy).
2. Increased heart rate and accelerated contraction (positive chronotropy).
3. An abbreviation of the Ca^{2+} transient and action potential duration, with a faster relaxation (positive lusitropy).

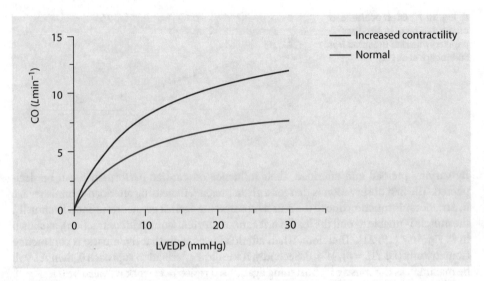

☐ **Fig. 10.8** Frank-starling curve at normal and increased contractility. Preload is represented by the LVEDP. Notice that increasing contractility (red curve) increases CO at any given preload

Fluctuations in energetic demands lead to corresponding changes in adrenergic stimulation to maintain the CO. As such, the heart is constantly exposed to varying concentrations of catecholamines, allowing it to operate on a whole spectrum of Frank–Starling curves proportional to the level of adrenergic stimulation.

10.1.5   Pressure–Volume Loops

We will now consider the effect of cardiac performance determinants in the context of pressure–volume (PV) loops. The heart is a pressure generator constantly expending energy to perform mechanical work. As we have seen at the level of ventricular strips or cardiomyocytes, this takes the form of myocyte shortening (ΔL) and force generation. At the level of the heart, mechanical work is best described by changes in ventricular pressure and volume. If a ventricle of area A (cm^2) contracts by length L (cm), then $L \times A$ represents a change in ventricular volume (ΔV, cm^3). As the ventricle is contracting, a force F is applied on A, such that F/A represents a change in pressure (ΔP) [8]. When the relationship between pressure and volume is plotted over an entire cardiac cycle, a PV loop is obtained [20].

In PV loops, volume is represented on the *x-axis*, whilst pressure is on the *y-axis*. Each side of the loop represents one phase of the cardiac cycle, drawn in counterclockwise directions beginning with

1. Isovolumetric contraction, whereby ventricular pressure rises at constant volume.
2. Rapid ejection phase, when pressure rises correspondent with a drop in ventricular volume.
3. Isovolumetric relaxation, during which pressure drops at constant volume.
4. Diastolic filling—associated with ventricular filling and small increases in volume and pressure.

◻ Fig. 10.9 Pressure-Volume Loop. Isovolumic contraction begins at point A, which denotes the end-diastolic volume and pressure. Ejection then follows, with point A′ denoting the end-systolic volume and pressure, followed by isovolumic relaxation. This is followed by diastolic filling. A and A′ are points on the end-diastolic pressure–volume (EDPVR) and end-systolic pressure–volume relationship (ESPVR), respectively

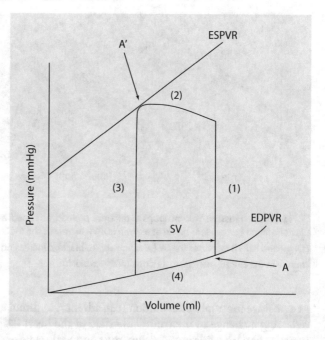

A typical PV loop is shown in ◻ Fig. 10.9. Stroke volume can be calculated as the difference between EDV and end-systolic volume (ESV) -that is, $SV = EDV - ESV$. Ejection fraction (EF), a measure of the proportion of blood ejected from EDV, can then be calculated [18] as $EF = SV/EDV$. In a healthy adult, stroke volume is ≈70 ml and EDV ≈ 120 ml, producing an EF of ≈0.58. EF is a measure of the pumping efficiency and is often altered in pathology.

The point on the PV loop just before isovolumic contraction (A) demarcates the end-diastolic volume and pressure, sitting on a given end-diastolic pressure–volume relationship (EDPVR). Likewise, the point at the end of the ejection phase (A′) denotes the end-systolic volume and pressure, sitting on the end-systolic pressure–volume relationship (ESPVR). Both EDPVR and ESPVR are important determinants of ventricular function. The EDV determines the stretch on myocardial fibres (and thus preload), with a higher EDV resulting in an increased SV (◻ Fig. 10.10a). The actual position of the EDV on the PV loop is dependent on the EDPVR, which describes the ventricular wall compliance. This can be altered in pathology, meaning that the same EDV may correspond to a different EDP if the EDPVR is altered [20]. As we have seen, a decrease in afterload allows the muscle to shorten more, decreasing ESV and thereby increasing SV for a given EDV (◻ Fig. 10.10b). In the same manner, as we have described the EDV and EDP at end-diastole, the same can be done for end-systole in a relationship known as ESPVR. The ESPVR tells us about the contractile state of the myocardium. Notice that for a change in afterload or preload (◻ Fig. 10.10a, b), the SV and the peak developed pressure (defined as the highest point on the PV loop) change, yet the end-systolic point always 'hits' the ESPVR line.

Thus, changes in loading of the heart (i.e. preload or afterload) may lead to changes in SV or peak pressure but will always end up on the ESPVR. This makes the ESPVR a *load-independent measure of contractility* [21]. In contrast, changes in the contractile state, such

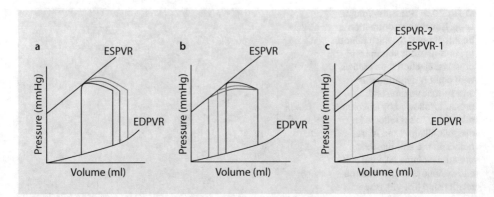

Fig. 10.10 Pressure-Volume loops at different preload, afterload, and inotropic states. **a** Increasing diastolic filling (green>red>black) at a given EDPVR increases the SV. **b** Decreasing afterload at a given EDPVR and ESPVR increases the SV (brown>purple>black). **c** Increasing the inotropic state causes a change in the ESPVR, increasing its slope (orange>black)

as a positive inotropic intervention (e.g. adrenergic stimulation), will change the ESPVR, reflecting a change in the contractile state of the heart (■ Fig. 10.10c). As the ESPVR is typically fit with a linear regression, the contractile state of the heart can be quantified by the slope of the ESPVR, such that a positive inotropic agent increases the slope, whereas a negative inotropic agent decreases it.

PV loops are the ultimate cardiac biomechanics descriptors, characterizing both the intrinsic properties of the ventricle and the coupling of the latter with the vasculature (e.g. increased afterload due to high blood pressure would shift the shape and height of a given PV loop). Together with technological advancements, they have enabled noninvasive assessment of ventricular properties, with immediate implications for diagnosis, prognosis and treatment of disease [22]. For example, with progressively worsening heart disease profile, it is widely accepted that EF typically falls. However, when an all-encompassing outcome including all-cause mortality, transplantation and mechanical circulatory support device implantation is correlated against PV loop parameters, the best predictor is EDV [23]. Likewise, changes in loop morphology can be used to appreciate ventricular remodelling as seen with progressive heart disease where both the width and the height of PV loops decrease, reflecting decreased SV and pressure generation, respectively [23].

10.2 Clinical Implications

The clinical importance of the regulators of myocardial contractility can be appreciated by investigating what happens when these go erratic. Heart failure (HF) is a variable clinical syndrome manifesting as the convergent endpoint pathway for a variety of pathological processes that impair myocardial function. One such process is arterial hypertension. High blood pressure is the single most important risk factor for HF, and it is estimated that up to 75% of HF cases have had antecedent hypertension [24]. In arterial hypertension, afterload is increased such that the ventricle typically spends a prolonged time in isometric contraction [25]. At such afterload the amount of work performed by the heart increases. To cope with this increased demand, the heart, being a terminally differentiated organ, enlarges leading to ventricular wall thickening (i.e. hypertrophy). If the pressure overload persists, this initially adaptive hypertrophic mechanism progresses towards maladaptive failure [24, 26]. When pathologically high blood pressure is modelled *in vivo* by chronic aortic band constriction, pathological ventricular remodelling ensues in the form of increased fibrosis, inflammation and apoptosis, coupled with decreased EF and aberrant calcium handling, ultimately progressing to HF [27]. Likewise, when preload is modelled *in vivo* by increasing ventricular volume via vena cava shunting progression to HF follows, yet the underlying pathways are not the same as those activated in response to pressure overload [27].

Pathological remodelling of the heart in response to such abnormal conditions of mechanical overload has been shown to be potentially reversible with the use of mechanical circulatory support devices, known as left-ventricular assist devices (LVADs) [28]. Data suggests that haemodynamic unloading mediates improvements in ventricular function and structure (i.e. 'reverse remodelling') by changes in myocyte size and structure, calcium homeostasis, extracellular matrix and fibrosis, as well as signal transduction pathways [28–30].

10.3 What We Don't Know and Where We are Heading: Isolating Mechanical Load

We have seen the importance of preload, afterload and contractility in determining the performance of the heart, their detrimental effects when they go haywire, as well as the clinical utility of PV loops in the assessment of heart physiology and pathology. However, a complete understanding of the full picture of mechanical loading on the function of the heart is still lacking. Abnormal preload and afterload both converge to heart dysfunction, yet their underlying pathways are not identical [27]. Additionally, although *in vivo* experiments [27] yield valuable information and are the final hurdles to be surpassed to progress to human trials, the haemodynamic system is intrinsically linked to the neurohormonal axis. In response to pressure overload or an injury to the heart, the drop in CO is met by a compensatory rise in the adrenergic system in order to maintain CO [31]. Despite its short-term beneficial effects, chronic adrenergic hyperactivation results in complex stimulation of signalling pathways eventually leading to deterioration of heart function [31]. This neurohormonal activation concomitant with the increased mechanical load makes *in vivo* models very complex systems with a huge matrix of variables to account for where separating the activation of molecular pathways due to hormonal stimulation from those of mechanical stimulation becomes extremely difficult. Basic science and intermediate models of myocardial physiology that can be studied *in vitro* but within a physiological environment of mechanical load and/or hormonal stimulation are the keys to understanding different and overlapping pathways of hormonal and mechanical systems and their interaction, and hold the promise for the development of much needed novel therapeutics.

References

1. Vincent JL (2008) Understanding cardiac output. Crit Care 12(4):174
2. Kumada M, Azuma T, Matsuda K (1967) The cardiac output-heart rate relationship under different conditions. Jpn J Physiol 17(5):538–555. [Internet] Available from: http://joi.jlc.jst.go.jp/JST.Journalarchive/jjphysiol1950/17.538?from=CrossRef
3. Wégria R, Frank CW, Wang H (1958) The effect of atrial and ventricular tachycardia on cardiac output, coronary blood flow and mean arterial pressure. Circ Res 6(5):624–632
4. Hamdani N, Kooij V, Van Dijk S, Merkus D, Paulus WJ, Dos RC et al (2008) Sarcomeric dysfunction in heart failure. Cardiovasc Res 77:649–658
5. Bers DM (2002) Cardiac excitation-contraction coupling. Nature 415:198–205
6. Sun YB, Irving M (2010) The molecular basis of the steep force-calcium relation in heart muscle. J Mol Cell Cardiol 48:859–865
7. Wiggers CJ (1951) Determinants of cardiac performance. Circulation 4(4):485–495
8. Levick JR (2009) An introduction to cardiovascular physiology, 5th edn. Hodder Arnold, London, pp 51–60
9. Bers DM (1991) Excitation-contraction coupling and cardiac contractile force. Springer, Netherlands
10. Allen DG, Kentish JC (1985) The cellular basis of the length-tension relation in cardiac muscle. J Mol Cell Cardiol 17:821–840
11. de Tombe PP, ter Keurs HEDJ (2016) Cardiac muscle mechanics: sarcomere length matters. J Mol Cell Cardiol 91:148–150
12. Fuchs F, Smith SH (2001) Calcium, cross-bridges, and the Frank-Starling relationship. News Physiol Sci 16(1):5–10. [Internet] Available from: http://physiologyonline.physiology.org/content/16/1/5.abstract
13. Solaro RJ (2007) Mechanisms of the Frank-Starling law of the heart: the beat goes on. Biophys J 93(12):4095–4096
14. Farman GP, Gore D, Allen E, Schoenfelt K, Irving TC, De Tombe PP (2011) Myosin head orientation: a structural determinant for the Frank-Starling relationship. Am J Physiol Heart Circ Physiol 300:2155–2160
15. Sonneblick HE, Downing SE (1963) Afterload as a primary determinant of ventricular performance. Am J Physiol Heart Circ Physiol 204:604–610
16. Nozawa T, Cheng CP, Noda T, Little WC (1994) Relation between left ventricular oxygen consumption and pressure-volume area in conscious dogs. Circulation 89(2):810–817
17. Suga H, Hayashi T, Shirahata M (1981) Ventricular systolic pressure-volume area as predictor of cardiac oxygen consumption. Am J Phys 240(1):H39–H44
18. Solaro RJ (2011) Regulation of cardiac contractility [internet]. In: Colloquium series on integrated systems physiology: from molecule to function, vol 3, pp 1–50. Available from: http://www.ncbi.nlm.nih.gov/books/NBK54078/
19. Florea VG, Cohn JN (2014) The autonomic nervous system and heart failure. Circ Res 114:1815–1826
20. Katz AM (1988) Influence of altered inotropy and lusitropy on ventricular pressure-volume loops. J Am Coll Cardiol 11(2):438–445
21. Suga H, Sagawa K, Shoukas AA (1973) Load independence of the instantaneous pressure-volume ratio of the canine left ventricle and effects of epinephrine and heart rate on the ratio. Circ Res 32:314
22. Ky B, French B, May Khan A, Plappert T, Wang A, Chirinos JA et al (2013) Ventricular-arterial coupling, remodeling, and prognosis in chronic heart failure. J Am Coll Cardiol 62:1165
23. Burkhoff D (2013) Pressure-volume loops in clinical research. J Am Coll Cardiol 62(13):1173–1176
24. Burchfield JS, Xie M, Hill JA (2013) Pathological ventricular remodeling: mechanisms: part 1 of 2. Circulation 128(4):388–400
25. Schotola H, Sossalla ST, Renner A, Gummert J, Danner BC, Schott P et al (2017) The contractile adaption to preload depends on the amount of afterload. ESC Hear Fail 4:468
26. Machackova J, Barta J, Dhalla NS (2006) Myofibrillar remodelling in cardiac hypertrophy, heart failure and cardiomyopathies. Can J Cardiol 22:953
27. Toischer K, Rokita AG, Unsöld B, Zhu W, Kararigas G, Sossalla S et al (2010) Differential cardiac remodeling in preload versus afterload. Circulation 122(10):993–1003

10

28. Drakos SG, Terrovitis JV, Anastasiou-Nana MI, Nanas JN (2007) Reverse remodeling during long-term mechanical unloading of the left ventricle. J Mol Cell Cardiol 43:231–242

29. Ibrahim M, Al Masri A, Navaratnarajah M, Siedlecka U, Soppa GK, Moshkov A et al (2010) Prolonged mechanical unloading affects cardiomyocyte excitation-contraction coupling, transverse-tubule structure, and the cell surface. FASEB J 24(9):3321–3329. [Internet] Available from: http://www.fasebj.org/cgi/doi/10.1096/fj.10-156638

30. Ibrahim M, Kukadia P, Siedlecka U, Cartledge JE, Navaratnarajah M, Tokar S et al (2012) Cardiomyocyte Ca2+handling and structure is regulated by degree and duration of mechanical load variation. J Cell Mol Med 16(12):2910–2918

31. Kishi T (2012) Heart failure as an autonomic nervous system dysfunction. J Cardiol 59:117–122

32. M. G. Hibberd, B. R. Jewell, (1982) Calcium- and length-dependent force production in rat ventricular muscle. The Journal of Physiology 329 (1):527-540

The Scientific Basis of Heart Failure

Giles Chick, Fotios G. Pitoulis, and Liam Couch

© Springer Nature Switzerland AG 2019
C. Terracciano, S. Guymer (eds.), *Heart of the Matter*, Learning Materials in Biosciences,
https://doi.org/10.1007/978-3-030-24219-0_11

What You Will Learn in This Chapter

The aim of this chapter is to discuss the current literature surrounding the science that underlies heart failure. This encompasses the natural history of the syndrome, specifically, the initial myocardial insult, apparent recovery, decompensation and finally failure of the organ. We will then expand upon some of the post-infarction changes to ventricular performance in both the compensatory and failure stages, including their manifestations on the Frank–Starling curve. Finally, the topic of cardiac regeneration will briefly be analysed. Understanding the science behind regeneration will provide a foundation to appreciate the basis of novel therapies, discussed later.

Learning Objectives

- Appreciate the global burden of heart failure, focussing particularly on why it is difficult to treat.
- Be able to provide a succinct overview of the four stages that define heart failure pathogenesis.
- Describe some of the mechanisms that collectively comprise compensatory cardiac remodelling.

11.1   Heart Failure: Evolution of a Definition

As our knowledge of heart failure (HF) has grown, its definition has evolved. The Egyptians, Greeks and Romans identified common manifestations of HF, including oedema, dyspnoea and cardiac hypertrophy. However, the first formal definition was formulated in 1933, when Thomas Lewis described it as a condition in which the heart fails to discharge its contents adequately [1]. This was made gradually more accurate, with Paul Wood describing it as a state in which the heart fails to maintain an adequate circulation for the needs of the body, despite a satisfactory filling pressure [1]. It wasn't until 1985 that it was first labelled as a syndrome by Philip Poole-Wilson and recognised as a pattern of haemodynamic, renal, neural and hormonal compensatory responses [1].

These definitions all categorised heart failure as a problem with systolic function, now termed *heart failure with reduced ejection fraction* (HFrEF). Notwithstanding this, we now understand that a cardiac failure phenotype can involve diastolic function, termed *heart failure with preserved ejection fraction* (HFpEF).

11.2   Why Is Heart Failure Important?

Heart failure in the UK is now described as an epidemic, with an estimated 1 million people in the UK suffering from the syndrome, and 120,000 new cases are diagnosed each year [2, 3]. The mortality rates of HF are similar to the most aggressive cancers, with 30–40% of patients dying within the first year of diagnosis and 10% annually thereafter [3]. Additionally, HF carries a high symptom burden; therefore, the quality of life management of each patient is critical and the need for an effective therapy is paramount.

With this in mind, it is unsurprising that the management of heart failure constitutes 1–2% of UK healthcare expenditure: a burden only expected to increase as the global population ages, with increasing survival rates for both acute myocardial infarctions (MI) and heart failure itself [2]. Historical therapeutic advancement has correlated with a decline in HF mortality, meaning it now makes the largest contribution to age-standardised years living with disability in high-income men in four continents of the world.

11.3 Heart Failure: The Natural History

The natural history of HF involves four key stages. In response to any one of a constellation of insults to the myocardium, apparent recovery or compensation occurs to restore the lost pumping capacity, mediated by chronic adrenergic activation and the renin–angiotensin–aldosterone system [4]. This acutely produces phenotypic improvement whilst longitudinally facilitating a period of decompensation once all compensatory mechanisms are exhausted. At this stage, the classical symptoms of dyspnoea, cyanosis and fatigue appear [4]. Once HF has decompensated, survival rates are poor.

11.3.1 Myocardial Insult

The commonest sources of myocardial damage are myocardial infarction (MI) and hypertension; however, valvular disease, genetic defects, arrhythmias, cardiomyopathies, alcohol and drugs (especially chemotherapeutic agents) also contribute to varying extents [5]. Any combination of these leads to a progressive deterioration in cardiac contraction, either via myocardial death or impaired cardiomyocyte function.

MIs are most commonly caused by the rupture of an atherosclerotic plaque inside a coronary artery (the pathophysiology of which shall be discussed in a later chapter), impairing blood flow to cardiac tissue supplied by the artery and resulting in ischaemia. This results in infarction and subsequent cardiac myocyte necrosis. Uncontrolled necrosis is characterised by the release of inflammatory cell contents, apoptotic mediators and further (potentially widespread) damage [6]. As a defence mechanism, the heart attempts to limit this through apoptosis. The cells most susceptible to ischaemic damage are located in the subendocardial region, with transmural necrosis characteristic of more significant infarction [7]. Fundamentally, the myocardium is then more likely to conduct abnormally due to scar formation, form aneurysms and become chronically inflamed [4]. Over time, these changes contribute towards the phenotype of the failing heart.

11.3.2 Compensation and Remodelling Mechanisms

To compensate for its reduced pumping capacity, the heart must adapt to magnify the capability of the remaining functional cardiomyocytes. Some of these measures include volume loading to maintain preload, hypertrophy to increase stroke volume (SV), dilation to enhance end-diastolic volume and chronically elevated sympathetic stimulation to maximise cardiac output and preserve blood pressure [4, 8]. In this way, compensation exists not only within the heart but systemically. A variety of ionic alterations also occur, which will be discussed in ▶ Chap. 13.

11.3.2.1 Impaired Contraction

The impairment of contraction exhibited in HF is best described using a well-known experiment conducted by Davies et al. (1995), whereby cardiomyocytes from a normal and failing heart were isolated, before being stimulated to varying degrees [9].

Pacing the cells at 0.2 Hz revealed similar contraction of both normal and failing cardiomyocytes, with the same amplitude of contraction. However, upon increased stimulation, failing myocytes exhibited slowing of contraction, increased time to maximal contraction and slower relaxation, a hallmark of failing hearts. Moreover, Ca^{2+} didn't return to baseline between contractions. Notably, loss of the Treppe effect was observed in the failing cells, a phenomenon whereby healthy cells increase their shortening (and thus contractility) at higher frequencies [9].

Cardiomyocytes from different disease states (mitral valve disease, ischaemia and idiopathic dilated cardiomyopathy) all had significantly slower relaxation times compared to a non-failing heart, suggesting a lack of functional reserve within the myocardium, that is to say, a reduced ability to modify functional capacity in response to increased demand [9].

11.3.2.2 Morphological Changes

Several alterations in cardiomyocyte morphology have been noted in failing hearts. Classically, this manifests as cells becoming wider and longer due to hypertrophy, with increased sarcomeric density and more defined branching patterns [10, 11]. It is arguable, however, that the role of individual morphological changes towards global functionality is less significant than other remodelling facets, since myocyte size has been demonstrated as poorly correlating with function [12]. When large myocytes from a normal heart were compared to normal-sized cells from a hypertrophic heart, the large cells from the healthy heart contracted normally whilst the small cells from the hypertrophied heart contracted abnormally [12, 13]. This derives from the fact that contraction amplitude and velocity were both reduced, and highlights that the impaired cardiomyocyte function relates to the diseased state itself, rather than directly to the size of the individual cell.

11.3.2.3 Increased Sympathetic Stimulation

The body's response to a failing heart is to increase sympathetic stimulation, principally to maintain cardiac output [4]. This is partly mediated through an increase in plasma noradrenaline (NA), which acts at vascular adrenergic receptors (AR) to increase total peripheral resistance and thus blood pressure in addition to exerting positive inotropic and chronotropic receptor signalling at the myocardium.

Paradoxical to these seemingly beneficial effects, NA is an independent predictor of mortality in heart failure, with higher levels associated with increased mortality [14, 15]. The chronicity of βAR stimulation by noradrenaline contributes to the desensitisation of the receptor over time, limiting the ability to respond to exercise and underlying the classical heart failure symptoms of breathlessness and fatigue [16].

Consequently, a higher concentration of noradrenaline is required in failing hearts to reach normal intracellular calcium concentration and contraction [16, 17]. This decrease in maximal response to sympathetic stimulation is homogenous amongst failing hearts but is related to the severity of the disease, the natural age-related decline in adrenoceptor response, and the aetiology of the disease itself [17]. Therefore, the relationship is not entirely clear due to confounding factors.

11.3.2.4 Neurohormonal Activation

In addition to increased sympathetic stimulation, several neurohormonal pathways are activated during the compensation phase. These primarily revolve around the renin–angiotensin–aldosterone system (RAAS), antidiuretic hormone (ADH, also known as *vasopressin*) and endothelin. This equilibrial shift enhances the production of compounds such as angiotensin II and aldosterone, both of which mediate a systematic elevation in blood pressure through vasoconstriction and the increasing of blood volume, the latter of which is achieved via hypernatraemia and the subsequent retention of water in the renal tubule [4].

In experiments where isolated cardiomyocytes are treated with a solution containing aldosterone and/or angiotensin II, a consistently direct and potent toxicity is observed, in addition to the interspersion of fibrosis [18, 19]. Notwithstanding the reparative replacement of necrotic with fibrosis tissue in any post-infarction response, the persistent upregulation of this pathway observed in heart failure ultimately impairs cardiac electrical conduction and thus contraction, as demonstrated in a rabbit model by Ogawa et al. [20].

Furthermore, a number of cardioprotective anti-proliferative mechanisms are depressed during the compensation phase, including the production of natriuretic peptide, bradykinin, nitric oxide and adrenomedullin. Whilst of lesser importance, this further shifts the equilibrium towards a deleterious phenotype that propagates (and indeed accelerates) the HF pathophysiology.

With many of these interplaying with the aforementioned RAAS system, ACE inhibitors (explained in a later chapter) have become a stalwart of HF therapeutics.

11.3.3 Decompensation

Briefly, decompensation of chronic HF often occurs from the exhaustion of compensatory mechanisms and is often initiated by systemic illness, such as infection, or by further insult to the myocardium [4, 17]. At this point, the salient clinical features of dyspnoea, peripheral oedema and fatigue become apparent.

11.3.4 Death

About half of patients die from progressive decompensated heart failure, and the remainder from sudden cardiac death precipitated by arrhythmia (interplaying with the deposition of fibrotic material) or an ischaemic event.

11.4 Cardiac Dysfunction: Systolic and Diastolic

Reductions in cardiac pumping capacity occur due to either systolic or diastolic dysfunction. In systolic dysfunction, the causes of which include myocardial infarction, dilated cardiomyopathy and viral myocarditis, there is reduced ejection of blood due to a diminished rate of pressure rise in isovolumetric contraction [21, 22]. In contrast, diastolic dysfunction is characterised by abnormal ventricular filling deriving from an insufficient rate of pressure drop in isovolumetric relaxation, in addition to increased diastolic stiffness (steeper EDPVR). Causes of diastolic dysfunction include hypertension, hypertrophic cardiomyopathies and restrictive diseases such as amyloidosis or pericarditis [21, 22].

Importantly though, this HF classification does not simply differentiate the syndrome according to the simple observation that systolic HF results from impaired ejection and diastolic HF results from impaired relaxation. This is because poor ejection leads to poor filling and vice versa, and as such, there are usually elements of both systolic and diastolic dysfunctions present, regardless of the initial cause of decline in pumping capacity [23].

Instead, they are differentiated based on ventricular architecture characteristics. In patients with systolic heart failure, volume overload deriving from renal retention of Na^+ and H_2O results in increased diastolic wall stress, causing lengthening of cardiomyocytes and the addition of sarcomeres in series to one another [23]. This is known as *eccentric hypertrophy*. Comparatively, diastolic HF is characterised by pressure overload, which serves to increases systolic wall stress, facilitating an increase in the cross-sectional area of cardiomyocytes. This is visible as transmural ventricular wall thickening, referred to as *concentric hypertrophy* [23].

In distinguishing between systolic and diastolic HF, ejection fraction (EF) is frequently used as a surrogate of ventricular performance. Accordingly, systolic HF is characterised as '*HF with reduced EF* (*HFrEF*)' and diastolic HF as '*HF with preserved EF* (*HFpEF*)'. Whilst SV is reduced in both, ejection fraction is disproportionately reduced in HFrEF due to impaired ejection, resulting in a relatively elevated end-diastolic volume [21]. In contrast, with the heart's diastolic capacity hindered in HFpEF, this presents as a reduced EDV, and normal ejection fraction. Furthermore, elevated afterload and reduced contractility both increase EDV in HFrEF, whilst reduced preload and reduced relaxation decrease EDV in HFpEF, further augmenting the disparity in ejection fraction.

11.4.1 Ventricular Performance in HFrEF and HFpEF

As discussed in the previous chapter, pressure–volume (P-V) loops are one way of assessing the mechanical work done by the left ventricle and its efficiency using real-time pressure and volume data. The x-axis demarcates LV volume and the y-axis represents intrachamber pressure. P-V loops are drawn in a counterclockwise direction.

◘ Figure 11.1 illustrates the P-V loop of a left ventricle with impaired contractility, observed in HFrEF. As shown, the ESPVR relationship is shifted down and to the right in HF (decrease in inotropic state), such that systole ceases at a greater blood volume (increased ESV). This reduces systolic emptying, causing (a) reduced SV (as less blood leaves the ventricle), and by extension (b) an increase in ventricular EDV and pressure (as more blood stays in the ventricles). The latter means that blood may exhibit retrograde flow, moving into the atria to increase atrial pressure. As there is no valve separating the atria with the pulmonary venous circulation, increased left atrial pressure results in pulmonary congestion, ultimately explaining why HF patients commonly experience dyspnoea.

As previously outlined, compensatory mechanisms acutely improve this impaired cardiac performance. The positive inotropy, mediated by adrenergic stimulation, shifts ESPVR upwards and to the left, increasing the ability of the heart to eject blood at a given EDV. Increased lusitropy shifts the EDPVR relationship downwards and to the right, which reduces the EDP, decreasing pulmonary pressures and congestion [17]. Simultaneously, vasoconstriction and elevated Na^+ retention both increase venous return

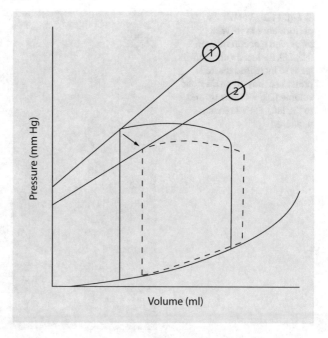

◘ Fig. 11.1 Ventricular performance in HF with reduced ejection fraction (HFrEF). A change in ESPVR relationship from normal [1] to abnormal [2] in HFrEF is visible

during the subsequent cardiac cycle, increasing EDV and thus preload, with a consequential increase in SV. Whilst initially helpful, chronic neurohormonal stimulation has detrimental consequences on ventricular performance. α-adrenoreceptor activation increases afterload, which is further augmented by the vasoconstrictive effects of Angiotensin II and vasopressin. This means that the acute compensatory CO increase is negated by increased afterload, causing increased aortic pressures, depressed SV and inefficient energy expenditure [4].

◘ Figure 11.2 illustrates a typical P-V loop and Frank–Starling curve for an HFpEF left ventricle. Increased diastolic stiffness (reduced compliance) causes the EDPVR in the P-V loop to shift upwards and to the left, hampering diastolic filling and preload. Yet, the ESPVR relationship is preserved such that the ventricle empties at the same ESV. Overall, this means that SV is reduced, however, with a fairly constant ejection fraction as the EDV is also reduced.

11.5 Co-morbidities in Heart Failure

Co-morbidities are common in heart failure, with Braunstein et al. observing that in a cohort of 122,630 patients, only 4% had no non-cardiac co-morbidities, with 39% having ≥5 and constituting 81% of the total inpatient hospital days [24]. Furthering this finding was van Deursen et al., who concluded from the European HF Pilot Study data that co-morbidities were independently associated with higher NYHA class ($p < 0.001$); however, only diabetes, CKD and anaemia independently increased hospitalisation risk [25]. Whilst this landmark study did highlight the specific importance of these three co-morbidities, both studies acknowledged the difficulties in distinguishing co-morbidities from potential aetiologies, with coronary artery disease, hypertension and arrhythmia just some of the aetiological panoply [24, 25].

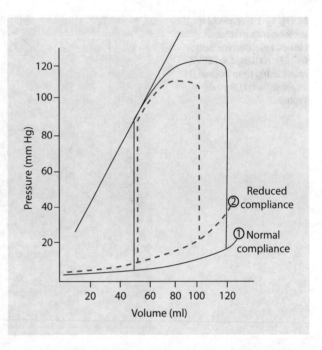

Fig. 11.2 Ventricular performance in HF with preserved ejection fraction (HFpEF). Reduced compliance leads to increased diastolic pressure at any given diastolic volume (EDPVR shifts up and to the left). ESPVR remains unaltered

The simultaneous management of a co-morbidity and HF is often challenging. This is exemplified by NSAIDS, which are commonly utilised for symptomatic relief and analgesia in arthritis, however, contraindicated in heart failure due to being associated with an increased risk of hospitalisation [26, 27]. Lacking this symptomatic management, Crofford reports that these co-morbid patients 'almost universally describe pain and stiffness', reducing both QOL and treatment adherence [27]. The physiological response to this inflammation involves adrenergic activation, the cardiac consequences of which include an increased arrhythmia risk, increased mechanical workload on the failing heart and cooperation in pathological remodelling, all of which contribute to deteriorating cardiac function [28].

11.6   What We Don't Know: Regeneration of the Failing Heart

Previously, heart failure was thought to centre around the heart's absent (or extremely limited to be precise) regenerative capacity. This remains a controversial area; however, although mitotic division has never been observed within the heart, some evidence of regeneration is provided by carbon dating. Carbon-14 emitted from nuclear weapons testing in the 1950s inadvertently resulted in the radiolabelling of cardiomyocyte nuclei in people alive during this period. By comparing the relative signal strength in cardiomyocytes to that expected, the estimated cell turnover could be calculated.

In this case, it revealed that the subjects' cardiomyocytes were labelled with different amounts of carbon-14 from the expected value, thus suggesting that some cardiomyocytes were produced after birth and that the heart possesses modest regenerative capacity [29]. Specifically, it was estimated that human cardiomyocytes have a turnover rate of 1% per year at 25, and 0.5% aged 75. Extrapolating this suggests that by the end of our life, we have a heart comprised of 50% cardiomyocytes from birth and 50% acquired throughout our life [29]. Whilst purely theoretical at this stage, this begs the notion that if the cardiomyocyte turnover rate could be increased in a manner that avoids the concomitant risk of cancer development, an entirely new therapeutic approach could be possible.

> **Take-Home Message**
>
> ▬ Heart failure is a clinical syndrome that follows four stages—myocardial insult, compensation, decompensation and death.
> ▬ Myocardial infarction and hypertension are the commonest initiating factors, with valvular disease, genetic defects, alcohol and drugs also being prominent aetiologies.
> ▬ Compensatory mechanisms involve both myocardial and systemic changes, which initially act to maintain cardiac output, however are deleterious in the long run.
> ▬ Techniques to regenerate damaged tissue are at the forefront of future therapies.

References

1. Davis RC (2000) ABC of heart failure: history and epidemiology. BMJ 320:297
2. Cowie MR (2017) The heart failure epidemic: a UK perspective. Echo Res Pract 4:R15
3. Bhatnagar P, Wickramasinghe K, Williams J, Rayner M, Townsend N (2015) The epidemiology of cardiovascular disease in the UK 2014. Heart 101:1182
4. Azevedo PS, Polegato BF, Minicucci MF, Paiva SAR, Zornoff LAM (2016) Cardiac remodeling: concepts, clinical impact, pathophysiological mechanisms and pharmacologic treatment. Arq Bras Cardiol 106(1):62–69
5. Pazos-López P, Peteiro-Vázquez J, Carcía-Campos A, García-Bueno L, de Torres JPA, Castro-Beiras A (2011) The causes, consequences, and treatment of left or right heart failure. Vasc Health Risk Manag 7:237–254
6. Frangogiannis NG (2014) The inflammatory response in myocardial injury, repair, and remodelling. Nat Rev Cardiol 11:255
7. Algranati D, Kassab GS, Lanir Y (2011) Why is the subendocardium more vulnerable to ischemia? A new paradigm. AJP Hear Circ Physiol 300:H1090
8. Burchfield JS, Xie M, Hill JA (2013) Pathological ventricular remodeling: mechanisms: part 1 of 2. Circulation 128:388
9. Davies CH, Davia K, Bennett JG, Pepper JR, Poole-Wilson PA, Harding SE (1995) Reduced contraction and altered frequency response of isolated ventricular myocytes from patients with heart failure. Circulation 92:2540
10. Gerdes AM, Kellerman SE, Moore JA, Muffly KE, Clark LC, Reaves PY et al (1992) Structural remodeling of cardiac myocytes in patients with ischemic cardiomyopathy. Circulation 86:426
11. Savinova OV, Gerdes AM (2012) Myocyte changes in heart failure. Heart Fail Clin 8:1
12. Botchway AN, Turner MA, Sheridan DJ, Flores NA, Fry CH (2003) Electrophysiological effects accompanying regression of left ventricular hypertrophy. Cardiovasc Res 60:510
13. Terracciano CMN, Hardy J, Birks EJ, Khaghani A, Banner NR, Yacoub MH (2004) Clinical recovery from end-stage heart failure using left-ventricular assist device and pharmacological therapy correlates with increased sarcoplasmic reticulum calcium content but not with regression of cellular hypertrophy. Circulation 109(19):2263–2265
14. Cohn JN, Tognoni G (2001) (Val-HeFT Trial) a randomized trial of the angiotensin-receptor blocker valsartan in chronic heart failure. N Engl J Med 345(23):1667–1675
15. Cohn JN, Johnson GR, Shabetai R, Loeb H, Tristani F, Rector T et al (1993) Ejection fraction, peak exercise oxygen consumption, cardiothoracic ratio, ventricular arrhythmias, and plasma norepinephrine

as determinants of prognosis in heart failure. The V-HeFT VA Cooperative Studies Group. Circulation 87(6 Suppl):VI5–V16

16. Bernstein D, Fajardo G, Zhao M (2011) The role of β-adrenergic receptors in heart failure: differential regulation of cardiotoxicity and cardioprotection. Prog Pediatr Cardiol 31:35

17. Madamanchi A (2007) β-adrenergic receptor signaling in cardiac function and heart failure. McGill J Med 10(2):99–104

18. He BJ, Joiner MLA, Singh MV, Luczak ED, Swaminathan PD, Koval OM et al (2011) Oxidation of CaMKII determines the cardiotoxic effects of aldosterone. Nat Med 17:1610

19. Tan LB, Jalil JE, Pick R, Janicki JS, Weber KT (1991) Cardiac myocyte necrosis induced by angiotensin II. Circ Res 69:1185

20. Ogawa M, Morita N, Tang L, Karagueuzian HS, Weiss JN, Lin SF et al (2009) Mechanisms of recurrent ventricular fibrillation in a rabbit model of pacing-induced heart failure. Hear Rhythm 6:784

21. Borlaug BA, Paulus WJ (2011) Heart failure with preserved ejection fraction: pathophysiology, diagnosis, and treatment. Eur Heart J 32(6):670–679

22. Nishimura RA, Jaber W (2007) Understanding "diastolic heart failure". J Am Coll Cardiol 49:695

23. Toischer K, Rokita AG, Unsöld B, Zhu W, Kararigas G, Sossalla S et al (2010) Differential cardiac remodeling in preload versus afterload. Circulation 122(10):993–1003

24. Braunstein JB, Anderson GF, Gerstenblith G, Weller W, Niefeld M, Herbert R et al (2003) Noncardiac comorbidity increases preventable hospitalizations and mortality among medicare beneficiaries with chronic heart failure. J Am Coll Cardiol 42(7):1226–1233

25. Van Deursen VM, Urso R, Laroche C, Damman K, Dahlström U, Tavazzi L et al (2014) Co-morbidities in patients with heart failure: an analysis of the European Heart Failure Pilot Survey. Eur J Heart Fail 16(1):103–111

26. Arfè A, Scotti L, Varas-Lorenzo C, Nicotra F, Zambon A, Kollhorst B et al (2016) Non-steroidal anti-inflammatory drugs and risk of heart failure in four European countries: nested case-control study. BMJ 354:i4857

27. Crofford LJ (2013) Use of NSAIDs in treating patients with arthritis. Arthritis Res Ther 15(Suppl 3):S2

28. Lymperopoulos A, Rengo G, Koch WJ (2013) Adrenergic nervous system in heart failure: pathophysiology and therapy. Circ Res 113(6):739–753

29. Bergmann O, Bhardwaj RD, Bernard S, Zdunek S, Barnabé-Heide F, Walsh S et al (2009) Evidence for cardiomyocyte renewal in humans. Science 324(5923):98–102

11

Molecular and Cellular Basis of Cardiomyopathies

Praveena Krishnakumar, Kabir Matwala, Shiv-Raj Sharma, and Salomon Narodden

© Springer Nature Switzerland AG 2019
C. Terracciano, S. Guymer (eds.), *Heart of the Matter*, Learning Materials in Biosciences,
https://doi.org/10.1007/978-3-030-24219-0_12

What You Will Learn in This Chapter

This chapter will begin by defining the term 'cardiomyopathy'. From there, a comprehensive overview of the five main cardiomyopathies will be provided, both discussing the clinical signs and symptoms it manifests as, and also the science that underlies these changes. Where relevant, specific genetic mutations will be explored. This chapter ties together themes from earlier in the book, including the arrhythmogenicity of macro-reentrant circuits, the concept of contractility and the cardiomyocyte microstructure. Finally, new research surrounding Takotsubo cardiomyopathy will be explored.

Learning Objectives

— Discuss the causes of hypertrophic cardiomyopathy and the phenotype this produces.
— Understand the distinct phenotypes of the five primary cardiomyopathies.
— Appreciate the role of desmosome protein mutations in the development of arrhythmogenic right ventricular cardiomyopathy (ARVC).

12.1 Cardiomyopathy: The Fundamentals

'*Cardiomyopathy*' refers to a collection of 'myocardial disorders in which the heart muscle is structurally and functionally abnormal in the absence of coronary artery disease, hypertension, valvular and congenital heart disease sufficient to cause the observed myocardial abnormality' [1]. First coined in 1952, this term came to refer three separate variants over the following decade, namely, *hypertrophic* (HCM), *dilated* (DCM) and *restrictive* (RCM) cardiomyopathies [2]. Subsequent imaging advances in the following resulted in the inclusion of further entities including *arrhythmogenic right ventricular* (ARVC) and so-called '*unclassified*' cardiomyopathies such as *Takotsubo*, also referred to as 'stress cardiomyopathy' and 'acute broken heart syndrome'. Importantly, with a prerequisite of this definition being both structural and functional abnormality, it therefore excludes channelopathies such as long QT and Brugada syndromes.

Of the five major categories of so-called 'primary cardiomyopathy' discussed, each can be classified into one of three major groups shown in ◘ Fig. 12.1, namely, genetic, acquired and mixed.

◘ **Fig. 12.1** Classification of the five main variants of primary cardiomyopathy

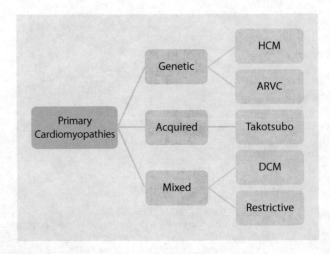

Catalysing the development of such categories has been an intensive effort to characterise genetic factors underlying the familial link of both the genetic and mixed cardiomyopathy phenotypes.

12.2 Hypertrophic Cardiomyopathy (HCM)

Hypertrophic cardiomyopathy (HCM), first described by Donald Teare in 1958, is characterised by cardiac hypertrophy independent of loading conditions, a non-dilated left ventricle (LV) and a normal or increased ejection fraction (EF) [3, 4]. It is mostly an autosomal disorder with variable penetrance [5]. HCM affects the sarcomere proteins and its associated protein, very rarely caused by autosomal recessive and X-linked modes of inheritance (e.g. Noonan syndrome [6], and Anderson–Fabry disease [7]. The commonest (30–50%) mutation is a missense (substitution) affecting gene *Myh7*, resulting in pathological alteration of ATPase activity and force generation of β-myosin heavy chains [8].

The second most common (20–40%) mutation affects gene *Mybpc3* and is a frameshift mutation caused by insertion/deletion that has phenotypic consequences on myosin-binding protein C [9]. A smaller proportion of patients (5–20%) exhibit mutations to *Tnnt2*, a gene encoding an essential component of the cardiac troponin T complex required for actomyosin interactions in response to Ca^{2+} [10]. Other less frequently affected genes are *Tpm1* (α-tropomyosin) and *Tnni3* (cardiac troponin I), both of which have an incidence of <5%.

As previously discussed, the classical HCM phenotype is asymmetrical hypertrophy of the LV, typically affecting the basal interventricular septum with LV outflow tract (LVOT) obstruction. Of course, other phenotypic variants of HCM also exist, namely, mid-cavity, concentric, apical and biventricular hypertrophy, in addition to concentric hypertrophic with cavity obliteration, and also progressive LV wall thinning [11–14].

Symptomatically, typical HCM manifestations include fatigue, dyspnea, presyncope or syncope, chest pain and palpitations, although patients may rarely be asymptomatic [7]. The presence of LVOT obstruction may be asymptomatic until the commencement of exercise. The majority of HCM patients also exhibit non-specific ECG abnormalities of hypertrophy, with a variable proportion having repolarisation abnormalities, abnormal Q-waves and inverted T-wave [8].

The echocardiographic hallmarks of HCM are the appearance of hypertrophy as described above, with potential haemodynamic features of LVOT obstruction and systolic anterior motion of the mitral valve [7]. Valsalva, exercise or pharmacological stressors such as dobutamine or nitrate-based agents could accentuate or unmask LVOT obstruction [15]. Histopathological features of HCM may include myocyte hypertrophy, an irregular 'chaotic' distribution of cells and unorthodox-shaped nuclei, and the presence of interstitial and replacement fibrosis. The intramural coronary arteries often appear with thickened vessel walls and a decreased lumen size [16]. A combination of these factors precipitates potential complications of heart failure (HF), sudden cardiac death (SCD) and stroke from an increased risk of atrial fibrillation [7].

12.3 Dilated Cardiomyopathy (DCM)

Dilated cardiomyopathy (DCM) is defined by the dual presence of LV dilatation and contractile dysfunction. It is a mixed cardiomyopathy, with acquired DCM constituting the majority of cases and deriving from a variety of aetiological factors. Up to 40% of cases are due to genetic factors, with over 40 causative genetic mutations identified [17]. These affect proteins of the sarcomere [18] (most frequently truncation mutations affecting titin [19], *Ttn*, but also myosin, *Myh6*, *Myh7*, *Mybpc3*; and actin, *Actc1* and *Actc2*), cytoskeleton [20] (desmin and cypher/ZASP), nuclear envelope [21] (limb-girdle muscular dystrophy, Emery–Dreifuss muscular dystrophy and autosomal dominant partial lipodystrophy), sarcolemma, ion channels and intercellular junctions. The latter two do not strictly belong to 'primary muscle disorders' to fit the definition of DCM.

In adult DCM, the usual mode of transmission is mostly autosomal dominant, commonly with partial and age-related penetrance and variable expression [22]. In paediatric and adolescent forms, autosomal recessive is the most common transmission pattern [23]. Phenotypically, DCM manifests as LV dilatation with impaired function (LVEF <40%) in the absence of hypertension, valvular heart disease and coronary heart disease [24]. The echocardiographic features of DCM include spherical dilatation of the LV, mitral regurgitation from annular dilatation and the potential appearance of pulmonary hypertension [24]. The global systolic cardiac function is reduced, and the diastolic function may show a restrictive filling pattern.

Patients with DCM may be asymptomatic for a number of years, with SCD an uncommon but notable first presentation of the disease. Typically, symptoms of HF occur over time, namely, a reduced exercise tolerance, dyspnoea and palpitations [25]. In contrast, the non-genetic (acquired) causes of DCM are diverse, some of which shown below in ▪ Table 12.1 [26–29].

12.4   Restrictive Cardiomyopathy (RCM)

Restrictive cardiomyopathy (RCM) is the least common of the three subtypes in the original 1952 definition and appears phenotypically as a non-dilated left or right ventricle with normal wall thickness [30]. Diastolic dysfunction may reveal as restrictive filling (due to

▪ **Table 12.1**   A summary of some notable non-genetic causes of DCM

Acquired causes of DCM	Examples
Infection	Adenovirus, coxsackie, HIV, cytomegalovirus, rheumatic fever, typhoid fever, syphilis, leptospirosis, histoplasmosis, toxoplasmosis
Prescribed drugs	Cyclophosphamide, zidovudine, clozapine
Toxins	Cocaine, alcohol, amphetamine, lead
Electrolyte imbalances	Uraemia and hypocalcaemia
Autoimmune pathology	SLE, scleroderma, dermatomyositis, rheumatoid arthritis
Endocrine pathology	Hypo/hyperthyroidism, gross hormone excess or deficiency, phaeochromocytoma

Information adapted from [26–29]

☐ Table 12.2 A summary of some notable causes of RCM

Familial causes of RCM	Examples
Endomyocardial	Sarcomeric protein mutations, troponin I (RCM +/− HCM), pseudoxanthoma elasticum, desminopathy
Infiltration	Essential myosin light chains and familial amyloidosis
Storage	Haemochromatosis, Anderson–Fabry disease, glycogen storage disease (type II, III, IV)
Non-familial causes of RCM	
Endomyocardial	Scleroderma, radiation, fibrosis-inducing drugs (e.g. serotonin, methysergide, ergotamine)
Infiltration	Metastatic cancers

Information adapted from [30–32]

decreased myocardial compliance) with elevated filling pressures and dilated atria. Systolic function is often preserved, albeit not completely normal [30]. Classical symptoms of RCM include dyspnoea, peripheral oedema, ascites, palpitations, fatigue, weakness and exercise intolerance. Diagnosis is usually made via exclusions. As a mixed cardiomyopathy, its cause in any given patient can be classified into familial (genetic) or non-familial (acquired), with the majority of cases being the latter [30]. Some of these are shown in ☐ Table 12.2 [30–32]. Mutations to genes encoding several sarcomeric proteins have been associated with RCM when inherited in an autosomal dominant manner, namely, β-myosin heavy chain (*Myh7*), actin (*Actc*), troponin I (*Tnni3*) and troponin T (*Tnnt2*) [33].

12.5 Arrhythmogenic Right Ventricular Cardiomyopathy (ARVC)

Arrhythmogenic right ventricular cardiomyopathy (ARVC) is an inherited disease characterised by the progressive replacement of myocardium by fibrofatty tissue in the RV [34]. Although non-ischaemic, the loss of myocardium results in regions of hypokinetic tissue, compromising the ventricle's contractile capability. Typically, this deposition of fibrofatty tissue is initially localised to three areas: the inflow tract, outflow tract and RV apex in an arrangement referred to as the 'triangle of dysplasia' [35]. Whilst predominating in the RV, advanced cases of ARVC frequently involve the left ventricle as well, in addition to some subvariants of ARVC concentrated in the LV [36]. ARVC manifests as symptoms including syncope and shortness of breath (dyspnoea), both of which derive from compromised RV contractility and thus output, in addition to palpitations.

Importantly, it is not uncommon for the first presentation of ARVC to be sudden cardiac death (SCD), particularly in young adults and competitive athletes [37]. In one study conducted over a 10-year period in Veneto, Italy, ARVC constituted the most frequently encountered cause of sudden cardiac death in young adults, accounting for 27% of 22 cases [37].

Potentially underlying the association between the symptomatic expression of ARVC (including SCD) and exercise is the increased myocardial workload and subsequent adrenergic signalling. This is superimposing on a tissue interspersed with fibrofatty tissue that will slow conduction, providing a suitable substrate for the generation of macro-reentrant circuits and ventricular arrhythmias [38].

ARVC is considered a genetic cardiomyopathy and is usually inherited in an autosomal dominant manner, with incomplete penetrance and a variable expression pattern. That said, some rare recessive cases have been noted [39]. Across several studies, causative mutations are typically identified in 30–50% of affected individuals, with a high proportion being in genes encoding desmosomal proteins [38]. It was only following this discovery that AVRC, originally considered a congenital defect of the right ventricle, was re-classified as a cardiomyopathy [38]. Desmosomes are a subset of junctional complex that confers stronger intercellular adhesion in order to maintain the global structural integrity of a tissue [40]. They are particularly abundant in tissues exposed to significant mechanical stress, such as the mucosa of the gastrointestinal tract, the epidermis and the myocardium, to name a few [40].

A desmosome comprises five structural elements, namely, plakoglobin, plakophilin-2, desmoplakin, desmoglein-2 and desmocollin-2. The encoding gene of each of the desmosome proteins is shown in ▢ Table 12.3.

In a seminal 2000 paper by McKoy et al., mutations in the gene encoding plakoglobin, *Jup*, were the first identified as contributing towards the development of ARVC [41]. This finding renewed interest on intercellular adhesion proteins, with subsequent causative mutations later identified in each of the genes encoding the other four proteins. The most frequently mutated of these is plakophilin-2, encoded by *Pkp2*, with one study of 120 unrelated ARVC individuals identifying heterozygous *Pkp2* mutations in 26.7% of cases [42].

The importance of desmosomes in the pathophysiology of ARVC is further supported by Basso et al. who used transmission electron microscope (TEM) to study the ultrastructure of intercalated discs obtained from ARVC patients. These were then compared against both control and DCM samples. The ARVC intercalated discs exhibited marked structural remodelling, including a reduction in the number of desmosomes and substantial elongation of those remaining. With desmosomes being the vital mechanical support structure of intercalated discs, a depletion in their representation represents a marked deterioration in the architectural stability of the heart as a functional syncytium. Specifically, with this reduction localised to the intercalated discs of the RV, a sudden increase in mechanical stress, such as that encountered during exercise, would represent a weakening of intercellular junctions in the ventricle.

▢ **Table 12.3**   Table of the five constituent desmosome proteins and their encoding genes

Desmosome protein	Encoding gene
Plakoglobin	*Jup*
Plakophilin-2	*Pkp2*
Desmoplakin	*Dsp*
Desmoglein-2	*Dsg2*
Desmocollin-2	*Dsc2*

12.6 Takotsubo Cardiomyopathy

Takotsubo cardiomyopathy (TCM) was first discovered in 1990 by Sato et al. and derives its name from the Japanese word '*takotsubo*', meaning 'octopus pot' [43]. This is due to the apical ballooning of the LV and resemblance this has to the aforementioned pot. Broadly, TCM has come to be characterised by transient and reversible systolic dysfunction concentrated in the middle and apical segments of the LV, with subsequent hypercontractility in the base of the heart [44].

Due to its association with physical and/or emotional stress, both of which frequently represent triggers of this cardiomyopathy, Takotsubo has also come to be known as 'stress cardiomyopathy' or 'acute broken heart syndrome' [43, 44]. Indeed, this is evidenced by Paur et al., who used an in vivo rat model to demonstrate the induction of mid-LV-to-apical hypocontractility (TCM) using high-dose IV adrenaline [44].

Likely underlying this distinctive phenotype is the heterogeneous distribution of β-adrenergic receptors (βAR) throughout the mammalian myocardium, with the apex characterised by the densest dispersion of receptors relative to the sparse proportion in the base [45, 46]. They form a broad βAR-response gradient that has been evidenced in several animal models [44, 47, 48].

Constituting ~90% of the total cardiac adrenergic receptors, βAR mediate many of the sympathetic nervous system (SNS) effects on the heart [49]. Briefly, catecholamines (especially adrenaline) bind to and activate βAR, triggering the exchange of GDP for GTP on the associated G_s alpha subunit. This facilitates dissociation of the G-protein, which is now able to activate adenylyl cyclase (AC), a 12-transmembrane domain enzyme that catalyses the conversion of ATP to cyclic 3′,5′-adenosine monophosphate (cAMP) and pyrophosphate using its C1a and C2a domains. Protein kinase A (PKA) is a holoenzyme composed of two regulatory and two catalytic subunits. cAMP binds to the two PKA regulatory subunits, causing dissociation from the catalytic subunits, which are now able to phosphorylate proteins, exerting many of the adrenergic effects, including positive inotropy (discussed further in ► Chap. 10).

However, $β_2AR$ is also able to exert signalling via an alternative G_i alpha subunit (and non-G-protein pathways altogether), with subsequent negative inotropic effects on the myocardium [44]. At high levels, adrenaline can 'switch' $β_2AR$ signalling to this pathway via the G_i alpha subunit in a process termed 'biased agonism'. Consequently, at high adrenaline concentrations, the differential apical–basal expression of $β_2AR$ produces intraventricular variability in inotropy and the Takotsubo phenotype [44]. For this reason, TCM is classified as an acquired cardiomyopathy.

Take-Home Message

- 'Cardiomyopathy' refers to myocardial disorders in which the heart is structurally and functionally abnormal in the absence of coronary artery disease, hypertension, valvular or congenital heart disease sufficient to cause the observed myocardial abnormality.
- Primary cardiomyopathies can be classified into genetic (HCM, ARVC), acquired (TCM) and mixed (DCM, RCM) categories.
- Arrhythmogenic right ventricular cardiomyopathy (ARVC) is characterised by the progressive replacement of myocardium by hypokinetic fibrofatty tissue, resulting in a loss of contractility in affected regions.

References

1. Elliott P, Andersson B, Arbustini E, Bilinska Z, Cecchi F, Charron P et al (2008) Classification of the cardiomyopathies: a position statement from the European Society of Cardiology Working Group on myocardial and pericardial diseases. Eur Heart J 29(2):270–276
2. Yacoub MH (2014) Decade in review – cardiomyopathies: cardiomyopathy on the move. Nat Rev Cardiol 11:628
3. Teare D (1958) Asymmetrical hypertrophy of the heart in young adults. Br Heart J 20(1):1–8
4. Marian AJ, Braunwald E (2017) Hypertrophic cardiomyopathy. Circ Res 121(7):749–770
5. Maron BJ (2002) Hypertrophic cardiomyopathy. JAMA 287(10):1308–1320
6. Hickey EJ, Mehta R, Elmi M, Asoh K, McCrindle BW, Williams WG et al (2011) Survival implications: hypertrophic cardiomyopathy in Noonan syndrome. Congenit Heart Dis 6:41
7. Semsarian C, Ingles J (2013) Expanding the genetic spectrum of hypertrophic cardiomyopathy: X marks the spot. Circ Cardiovasc Genet 6:528
8. Bonne G, Carrier L, Richard P, Hainque B, Schwartz K (1998) Familial hypertrophic cardiomyopathy: from mutations to functional defects. Circ Res 83:580
9. Maron BJ, Maron MS, Semsarian C (2012) Genetics of hypertrophic cardiomyopathy after 20 years: clinical perspectives. J Am Coll Cardiol 60:705
10. Richard P, Charron P, Carrier L, Ledeuil C, Cheav T, Pichereau C et al (2003) Hypertrophic cardiomyopathy: distribution of disease genes, spectrum of mutations, and implications for a molecular diagnosis strategy. Circulation 107:2227
11. Kim EK, Lee SC, Hwang JW, Chang SA, Park SJ, On YK et al (2016) Differences in apical and non-apical types of hypertrophic cardiomyopathy: a prospective analysis of clinical, echocardiographic, and cardiac magnetic resonance findings and outcome from 350 patients. Eur Heart J Cardiovasc Imaging 17:678
12. Bejiqi R, Retkoceri R, Bejiqi H (2011) Hypertrophic cardiomyopathy associated with mid-cavity obstruction and high left intraventricular pressure. Acta Inform Medica 19(4):241
13. Yamaguchi H, Ishimura T, Nishiyama S, Nagasaki F, Nakanishi S, Takatsu F et al (1979) Hypertrophic nonobstructive cardiomyopathy with giant negative T waves (apical hypertrophy): ventriculographic and echocardiographic features in 30 patients. Am J Cardiol 44(3):401–412
14. Olivotto I, Cecchi F, Poggesi C, Yacoub MH (2012) Patterns of disease progression in hypertrophic cardiomyopathy an individualized approach to clinical staging. Circ Hear Fail 5:535
15. Maron MS, Olivotto I, Zenovich AG, Link MS, Pandian NG, Kuvin JT et al (2006) Hypertrophic cardiomyopathy is predominantly a disease of left ventricular outflow tract obstruction. Circulation 114:2232
16. Hughes SE (2004) The pathology of hypertrophic cardiomyopathy. Histopathology 44:412
17. Mestroni L, Brun F, Spezzacatene A, Sinagra G, Taylor MR (2014) Genetic causes of dilated cardiomyopathy. Prog Pediatr Cardiol 37(1–2):13–18
18. Kamisago M, Sharma SD, DePalma SR, Solomon S, Sharma P, McDonough B et al (2000) Mutations in sarcomere protein genes as a cause of dilated cardiomyopathy. N Engl J Med 343:1688
19. Herman DS, Lam L, Taylor MRG, Wang L, Teekakirikul P, Christodoulou D et al (2012) Truncations of titin causing dilated cardiomyopathy. N Engl J Med 366:619
20. Sequeira V, Nijenkamp LLAM, Regan JA, Van Der Velden J (2014) The physiological role of cardiac cytoskeleton and its alterations in heart failure. Biochim Biophys Acta Biomembr 1838:700
21. Dellefave L, McNally EM (2010) The genetics of dilated cardiomyopathy. Curr Opin Cardiol 25:198
22. Hershberger RE, Morales A (1993) Dilated cardiomyopathy overview [Internet]. GeneReviews®. University of Washington, Seattle. [cited 2019 Jan 29]. Available from: http://www.ncbi.nlm.nih.gov/pubmed/20301486
23. Mestroni L, Rocco C, Gregori D, Sinagra G, Di LA, Miocic S et al (1999) Familial dilated cardiomyopathy: evidence for genetic and phenotypic heterogeneity. Heart Muscle Disease Study Group. J Am Coll Cardiol 34:181
24. Thomas DE, Wheeler R, Yousef ZR, Masani ND (2009) The role of echocardiography in guiding management in dilated cardiomyopathy. Eur J Echocardiogr 10:iii15
25. Michels VV, Driscoll DJ, Miller FA, Olson TM, Atkinson EJ, Olswold CL et al (2003) Progression of familial and non-familial dilated cardiomyopathy: long term follow up. Heart 89:757

12

26. Felker GM, Thompson RE, Hare JM, Hruban RH, Clemetson DE, Howard DL et al (2000) Underlying causes and long-term survival in patients with initially unexplained cardiomyopathy. N Engl J Med 342:1077

27. Pinto YM, Elliott PM, Arbustini E, Adler Y, Anastasakis A, Böhm M et al (2016) Proposal for a revised definition of dilated cardiomyopathy, hypokinetic non-dilated cardiomyopathy, and its implications for clinical practice: a position statement of the ESC working group on myocardial and pericardial diseases. Eur Heart J 37:1850

28. San Martín MA, García A, Rodríguez FJ, Terol I (2002) Dilated cardiomyopathy and autoimmunity: an overview of current knowledge and perspectives. Rev Esp Cardiol 55(5):514–524

29. Schürer S, Klingel K, Sandri M, Majunke N, Besler C, Kandolf R et al (2017) Clinical characteristics, histopathological features, and clinical outcome of methamphetamine-associated cardiomyopathy. JACC Hear Fail 5:435

30. Muchtar E, Blauwet LA, Gertz MA (2017) Restrictive cardiomyopathy: genetics, pathogenesis, clinical manifestations, diagnosis, and therapy. Circ Res 121:819

31. Kaski JP, Syrris P, Burch M, Tomé Esteban MT, Fenton M, Christiansen M et al (2008) Idiopathic restrictive cardiomyopathy in children is caused by mutations in cardiac sarcomere protein genes. Heart 94:1478

32. Navarro-Lopez F, Llorian A, Ferrer-Roca O, Betriu A, Sanz G (1980) Restrictive cardiomyopathy in pseudoxanthoma elasticum. Chest 78:113

33. Towbin JA (2014) Inherited cardiomyopathies. Circ J 78(10):2347–2356

34. Thiene G, Nava A, Corrado D, Rossi L, Pennelli N (1988) Right ventricular cardiomyopathy and sudden death in young people. N Engl J Med 318:129

35. Basso C, Thiene G, Corrado D, Angelini A, Nava A, Valente M (1996) Arrhythmogenic right ventricular cardiomyopathy: dysplasia, dystrophy, or myocarditis? Circulation 94:983

36. Norman M, Simpson M, Mogensen J, Shaw A, Hughes S, Syrris P et al (2005) Novel mutation in desmoplakin causes arrhythmogenic left ventricular cardiomyopathy. Circulation 112:636

37. Corrado D, Thiene G, Nava A, Rossi L, Pennelli N (1990) Sudden death in young competitive athletes: clinicopathologic correlations in 22 cases. Am J Med 89:588

38. Lombardi R, Marian AJ (2010) Arrhythmogenic right ventricular cardiomyopathy is a disease of cardiac stem cells. Curr Opin Cardiol 25:222

39. Marcus FI, Edson S, Towbin JA (2013) Genetics of arrhythmogenic right ventricular cardiomyopathy: a practical guide for physicians. J Am Coll Cardiol 61:1945

40. Garrod D, Chidgey M (2008) Desmosome structure, composition and function. Biochim Biophys Acta Biomembr 1778:572

41. McKoy G, Protonotarios N, Crosby A, Tsatsopoulou A, Anastasakis A, Coonar A et al (2000) Identification of a deletion in plakoglobin in arrhythmogenic right ventricular cardiomyopathy with palmoplantar keratoderma and woolly hair (Naxos disease). Lancet 355:2119

42. Gerull B, Heuser A, Wichter T, Paul M, Basson CT, McDermott DA et al (2004) Mutations in the desmosomal protein plakophilin-2 are common in arrhythmogenic right ventricular cardiomyopathy. Nat Genet 36:1162

43. Templin C, Ghadri JR, Diekmann J, Napp LC, Bataiosu DR, Jaguszewski M et al (2015) Clinical features and outcomes of Takotsubo (stress) cardiomyopathy. N Engl J Med 373:929

44. Paur H, Wright PT, Sikkel MB, Tranter MH, Mansfield C, O'Gara P et al (2012) High levels of circulating epinephrine trigger apical cardiodepression in a β2-adrenergic receptor/Gi-dependent manner: a new model of takotsubo cardiomyopathy. Circulation 126:697

45. Mori H, Ishikawa S, Kojima S, Hayashi J, Watanabe Y, Hoffman JIE et al (1993) Increased responsiveness of left ventricular apical myocardium to adrenergic stimuli. Cardiovasc Res 27:192

46. Lyon AR, Rees PSC, Prasad S, Poole-Wilson PA, Harding SE (2008) Stress (Takotsubo) cardiomyopathy – a novel pathophysiological hypothesis to explain catecholamine-induced acute myocardial stunning. Nat Clin Pract Cardiovasc Med 5:22

47. Mantravadi R, Gabris B, Liu T, Choi BR, De Groat WC, Ng GA et al (2007) Autonomic nerve stimulation reverses ventricular repolarization sequence in rabbit hearts. Circ Res 100:e72

48. Lathers CM, Levin RM, Spivey WH (1986) Regional distribution of myocardial beta-adrenoceptors in the cat. Eur J Pharmacol 130(1–2):111–117

49. O'Connell TD, Jensen BC, Baker AJ, Simpson PC (2014) Cardiac alpha1-adrenergic receptors: novel aspects of expression, signaling mechanisms, physiologic function, and clinical importance. Pharmacol Rev 66(1):308–333

Substrate Remodeling in Heart Failure

Angeliki Iakovou, Samuel Guymer, and Rasheda Chowdhury

© Springer Nature Switzerland AG 2019
C. Terracciano, S. Guymer (eds.), *Heart of the Matter*, Learning Materials in Biosciences,
https://doi.org/10.1007/978-3-030-24219-0_13

What You Will Learn in This Chapter

With the fundamentals of heart failure already discussed, this chapter will explain three integral facets of the remodeling process: electrophysiological, fibrotic, and mitochondrial. Instigated by interacting networks of neurohormonal and chemical signaling, these systems collectively contribute to much of the longitudinal functional decline observed, as well as significantly heightening the risk of sudden cardiac death. In addition to exploring their alteration in heart failure, this chapter will provide a brief overview of the fundamental pathway that underlies angiotensin-II-mediated fibrosis. Finally, novel pharmacological agents to combat these deleterious changes will be evaluated.

Learning Objectives

- Describe the changes to individual currents in heart failure, and the effect these have on global Ca^{2+} handling.
- Evaluate the role of fibroblasts and the extracellular matrix in post-myocardial infarction substrate remodeling and the ramifications this has on arrhythmogenesis.
- Summarize reactive oxygen species (ROS)-induced ROS release and explain the rationale underlying elamipretide activity.

13.1 Electrophysiological Remodeling

Advances in treatment options has precipitated a prolongation in the life expectancy of patients both acutely after myocardial infarction (MI) and following the diagnosis of heart failure (HF). However, sudden cardiac death (SCD) still represents a significant proportion of cases of cardiovascular death within HF patients, with the most common etiologies relating to aberrant ventricular electrical activity and the development of ventricular fibrillation (VF) [1]. As previously discussed, the ventricular action potential (AP) is the convergent electrical profile generated by a cluster of interdependent ionic currents whose magnitude varies over time (◘ Fig. 13.1).

In the acute and long-term phases following MI, the alteration in individual currents and the consequent action this has on other ionic fluxes serve to pathologically re-shape the ventricular action potential, heightening the risk of re-entrant circuits and other arrhythmogenic mechanisms [2]. This remodeling occurs in both the atrial and ventricular myocardium. Whilst some gaps in our understanding persist, considerable experimental data have identified several distinctive changes that tend to occur with fairly high frequency following MI [2–4]. These changes, in the expression, density, and function of ion channels, are accompanied by maladaptive cell death, myocyte apoptosis, and hyperplasia of the ECM, and collectively instigate a prolongation of the action potential duration [3, 5]. This is summarized as follows:

- Decreased Na^+/K^+ ATPase current
- Increased NCX current
- Increased persistent Na^+ current
- Decreased K^+ (repolarizing) currents

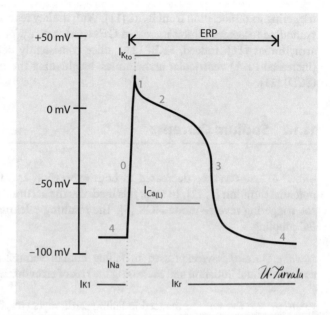

☐ Fig. 13.1 Illustration of a typical ventricular action potential. Each of the five phases possesses a unique ionic profile based on the transient activation of ion transporters

13.1.1 Calcium Currents

L-type Ca^{2+} Channel Current (I_{Ca-L}) normally triggers the release of Ca^{2+} from the sarcoplasmic reticulum (SR), with the density of this current corresponding to the magnitude of calcium-induced calcium release (CICR). With this current reduced in heart failure, a reduction in SR Ca^{2+} release is observed, resulting in weaker contractions [6].

Sarco/Endoplasmic Reticulum Ca^{2+}-ATPase (SERCA) reduces expression and function of *SERCA2a* (the main human cardiac isoform) in heart failure, lessening the uptake of Ca^{2+} into the SR to facilitate defective sequestration [7, 8]. Consequently, the amplitude and rate of decay of the calcium transient are both decreased, meaning slower relaxation and less efficient subsequent ejections [7].

Clinically, because phospholamban (PLB) inhibits SERCA2a function when dephosphorylated, phosphorylation of this protein could facilitate restoration of SERCA2a function in HF. In contrast to this, in a cardiomyocyte model, the expression of phospholamban has been successfully disrupted via the transfection of PLB antisense RNA via a recombinant adeno-associated virus (rAAV) vector, resulting in the upregulation of SERCA activity and thus reducing resting cytoplasmic [Ca^{2+}] [9].

Sodium-Calcium Exchanger (NCX) upregulated, both at an mRNA and a protein level, possibly to compensate for defective SERCA function [10]. Enhancement of this current drives a net inward +1 current flux, promoting depolarization of the cardiomyocyte and development of arrhythmias [10].

FKBP12.6 HF is characterized by a reduction in SERCA activity, which along with RyR hyperphosphorylation, increases Ca^{2+} leakage from the sarcoplasmic reticulum [8]. Normally, RyR2 is stabilized by a regulatory protein, FKBP12.6. Wehrens et al. characterized the aberrant RyR activity in HF as being driven by hyperphosphorylation of FKBP12.6, triggering its dissociation from RyR2 [11]. With the key regulatory apparatus now lacking, ryanodine receptors exhibit increased Ca^{2+} spark activity, promoting the development of arrhythmias [11]. Indeed, FKBP12.6$^-$ mice consistently demonstrate exercise-induced (increased PKA) ventricular arrhythmias, heightening the risk of sudden cardiac death (SCD) [11].

13.1.2 Sodium Currents

Na^+/K^+ ATPase Current decreased in both expression and function, prolonging action potential duration [2, 12]. In turn, this mediates the accumulation of calcium in the cytosol, triggering reverse-mode NCX [3]. The resulting calcium influx further dysregulates EC coupling.

I_f(Funny Current) overexpressed in failing human ventricles, thus promoting aberrant action potential initiation and increasing the risk of arrhythmias formation [13].

Persistent Na^+ Current enhanced in failing cardiomyocytes, furthering the augmenting of cytosolic sodium to propel the upregulated NCX activity.

13.1.3 Potassium Currents

Transient Outward Current (I_{to}) activated shortly after depolarization and responsible for the notch of repolarization. Despite some regional differences in the two primary genes encoding I_{to} (Kv$_{1.4}$ and Kv$_{4.3}$), the repolarizing current experiences an overall decrease in magnitude, which is demonstrated in canine studies [4]. This reduction in phase 1 of the ventricular action potential subsequently lessens the cellular Ca^{2+} influx via L-type Ca^{2+} channels (LTCCs), seemingly paradoxical to the overall AP prolongation observed in heart failure.

Inward Rectifier Current (I_{K1}) normally responsible for the return to and maintenance of resting membrane potential. I_{K1} channel density is significantly reduced in HF, which further contributes to the prolongation of action potentials [14]. Moreover, the reduction in current magnitude enhances automaticity in the failing heart, with Kurata et al. (2005) successfully converting fast response myocytes to cells exhibiting automaticity through the genetic suppression of I_{K1} [15].

Delayed Rectifier Currents a reduction in the outward current during the plateau range of voltages has been demonstrated to promote the development of arrhythmogenic early afterdepolarizations (EADs) via the prolongation of action potential duration.
There are two primary effects of the overall reduction in potassium currents:
1. Action potential duration is increased
2. Decreased I_{K1} reduces resting membrane potential stability → increases automaticity

13.1.4   Summary of Integrated Ca^{2+} Changes

The reduction in SERCA activity diminishes the sequestration of Ca^{2+} back into the SR, leading to an accumulation within the cytosol. A partial compensatory mechanism for this is the upregulation of NCX, effluxing one Ca^{2+} ion out of the cell in exchange for the influx of three Na$^+$ ions. Collectively, this translates to a net +1 inward current, promoting transient depolarizations and the development of delayed afterdepolarizations (DADs) [2, 3]. These changes in Ca^{2+} transients can be observed experimentally, where failing hearts exhibit smaller Ca^{2+} transients, with a decelerated repolarization phase [16].

13.2   Fibrosis Remodeling

As we have already seen, fibrosis remodeling represents an integral, yet potentially reversible, pathogenic step in the development of MI-mediated heart failure. Before we discuss the extracellular matrix (ECM) changes characteristic of heart failure and the architectural and electrical ramifications these have, it is important to first discuss the key mediators involved.

13.2.1   The Fundamentals

The ECM is a broad, highly organized, three-dimensional nexus comprising structural proteins, enzymes, adhesive glycoproteins, and protein-polysaccharide complexes.

13.2.1.1   Fibroblasts

Fibroblasts are the most abundant cell type by number and play a vital role in the production and maintenance of the ECM, both through the synthesis of structural components and the release of degradative factors such as matrix metalloproteinases (MMPs) [17]. Moreover, fibroblasts produce a number of active peptides including cytokines and growth factors in order to influence the hypertrophy and apoptosis of neighboring myocytes [17]. Fibroblasts are nonexcitable; however, they do possess ion channels and thus may facilitate the conduction of action potentials.

13.2.1.2   Collagen

Collagen is a structural protein in the ECM, which dually acts as a scaffold for the force of contraction whilst increasing the global elasticity of the tissue. If myocytes were merely joined by fascia adherens, they would separate and tear under force. In reality, myocytes are connected via collagen in addition to other intercellular machinery, creating uniform and synchronized contractions. Collagen is synthesized by fibroblasts as a pro-α chain comprising N-amino and carboxy terminal propeptide sequences, which flank a series of Gly-X-Y repeats [18]. Three pro-α chains are joined forming a triple helix, while the termini propeptides are enzymatically cleaved [18]. Different α-chain combinations then give rise to different collagen variants.

Several types of collagen exist in different regions of the heart, according to the tissue stiffness and contractility requirements. The most important types are collagens I and III, which are critical to the overall left ventricular pump function by coordinating myocyte shortening [19]. Collagen I is composed of two α1 chains and one α2 chain arranged in a tight right-handed triple helix [18]. Upon excretion into the ECM, it is enzymatically cleaved to produce a stable structure with a half-life of 80–120 days, partly explaining the slow turnover of scar tissue. Type I collagen is thick, therefore providing crucial tensile strength to resist mechanical stress. Conversely, type III collagen fibers are thinner in nature, conferring elasticity to the myocardium.

13.2.1.3 Adhesive Glycoproteins

Glycoproteins can form bridges between structural ECM molecules to reinforce the network within the ECM whilst connecting the network to soluble and cellular structures [20]. One noteworthy glycoprotein is fibronectin, which plays a vital role in the adhesion and migration of cells. Another glycoprotein, laminin, is extensively present throughout the heart, which is synthesized by cardiomyocytes, vascular smooth muscle cells, and endothelial cells.

13.2.1.4 Protein-Polysaccharide Complexes

Proteoglycans form the basis of higher order ECM structures around cells and are composed of a core protein covalently linked to glycosaminoglycans (GAGs): long chains of negatively charged disaccharide repeats [20]. GAGs confer resistance to compressive forces and bind to water to provide hydration. Major GAGs include heparan sulfate and chondroitin sulfate.

13.2.2 Fibroblasts in Disease

Whilst fibroblasts are critical to normal cardiac physiology, their role can become pathological in response to cardiac injury or infarction. Fibroblasts have two primary developmental origins: from epicardial cells via epicardial-to-mesenchymal transformation (EMT) and from endocardial cells via endocardial-to-mesenchymal transformation (EndMT) [17]. However, following myocardial injury, these two lineage pathways become activated once again in order to replace lost fibroblasts, in addition to hemopoietic stem cells in the bone marrow, which are recruited to the injury site and similarly differentiate into neo-fibroblasts [17].

Moreover, some disease states are characterized by the simultaneous proliferation and transition of fibroblasts into a myofibroblastic phenotype, especially in response to myocardial injury [21]. Myofibroblasts are considered to have greater mobility and contractile properties than fibroblasts, with an enhanced synthetic capacity to produce ECM proteins. In this sense, despite a loss of myocytes, myofibroblasts are able to replace some of the lost contractile behavior, unlike fibroblasts. Interestingly, myofibroblasts express a number of smooth muscle markers not typically expressed in inactive fibroblasts, such as α-smooth muscle actin (α-SMA) and fibronexus adhesion complexes, which bind the internal microfilament network to extracellular fibronectin, increasing the ECM contractile force [21].

13

13.2.2.1 Aberrant ECM Regulation in Disease

Whilst normally assisting in the maintenance of homeostasis, the ECM also responds to stress or injury. Indeed, these stress signals can stimulate remodeling of the ECM, facilitating healthy recovery or leading to further complications and heart failure. This shift is mediated in part by an upregulation in the enzymatic activity of matrix metalloproteinases (MMPs) and corresponding decrease in the activity of tissue inhibitors of metalloproteinases (TIMPs) [22]. Whilst appearing detrimental to the ECM architectural integrity, this is counteracted by an increase in fibroblast proliferation following injury, with escalating deposition of collagen and excretion of profibrotic factors [22]. The excessive production of ECM proteins serves to increase myocardial pressure, further exacerbating cardiac dysfunction.

13.2.2.2 Arrhythmogenic Properties of Fibrosis

As previously mentioned, fibroblasts do not exhibit action potentials; however, they can transfer current [23]. Importantly, they also have a "less negative" resting membrane potential compared to cardiomyocytes (CMs). Due to the presence of gap junctions between fibroblasts and CMswhich act as an electrical connection, fibroblasts can modulate electrical activity by "pulling up" the resting potential of neighboring cardiomyocytes via coupling. This leads to an increased probability that they will reach the depolarizing threshold, namely the activation threshold of voltage-gated Na^+ channels [23, 24]. Similarly, if a fibroblast is at a lower potential than its coupled CMs, it acts to draw down the voltage of neighboring cells to match its own, accelerating repolarization relative to if the fibroblast was not there. In this way, if repolarization occurs more quickly, there is a shorter AP duration and an increased risk of re-entry [24].

At an organ level, upregulated fibrosis results in the formation of electrically inactive regions, isolating strands of myocardium. Consequently, an excitatory wavefront is forced to follow more convoluted paths, increasing the likelihood of re-entry [25]. Each of the four main categories of tissue fibrosis (interstitial, diffuse, patchy, and compact) possesses a different arrhythmogenic profile, with compact fibrosis being the least arrhythmogenic [25]. Whilst macro re-entry can still occur in compact, it is characterized by large areas of densely packed collagen that hinder the potential for the deceleration and/or unidirectional block of current. The greatest risk arises from interstitial and patchy fibrosis, each of which separates myocyte bundles to force inter-myocyte connectivity via slower patterns of "zigzag" conduction [25].

13.2.3 Ang-II-Mediated Upregulation of Fibrosis

We have previously heard how fibrosis remodeling represents an integral facet of the HF pathophysiology. Essential to this is the upregulation of Ang-II signaling, deriving from chronic activation of the β-adrenergic and renin-angiotensin-aldosterone (RAAS) systems, especially following myocardial infarction [26]. Broadly speaking, post-infarction fibrosis remodeling can be classified into two stages that form a biphasic longitudinal relationship. The first of these is *replacement fibrosis*, which occurs shortly after MI and is necessary for the spatially localized deposition of (mostly type 1) collagen and ECM proteins to replace necrosed myocytes in the infarcted ventricular wall [27]. This is crucial to prevent hemorrhage into the pericardium.

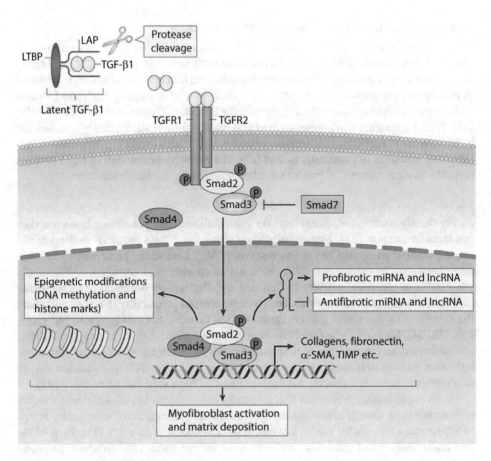

Fig. 13.2 Ang-II-TGF-β1 fibrosis pathway, with TGF-β1 binding to TGF-β receptor-2 (TGFβR2) on the plasma membrane and TGFR1 recruitment triggering a cascade of Smad-dependent-and-independent signaling. Smad2 and 3 complexes with Smad4 upregulate profibrotic pathways. (Reproduced with permission from Meng et al. [26])

13

As remodeling progresses, a multifaceted signaling nexus instigates a transition into the less upregulated but more spatially indiscriminate *reactive fibrosis*, characterized by the aberrant deposition of collagen in noninfarcted myocardium [27]. Underlying this is a central Ang-II-TGF-β1 canonical signaling pathway, as illustrated in Fig. 13.2. The binding of TGF-β1 to TGF-β receptor-2 (TGFβR2) on the plasma membrane triggers the recruitment and transphosphorylation of the type 1 (TGFβR1) receptor [26]. This in turn activates a nexus of Smad-dependent and Smad-independent transduction pathways, with Smad2/3 transcription factors forming complexes with Smad4 to upregulate profibrotic expression programs, including the transdifferentiation of fibroblasts to myofibroblasts [26].

Along with hemodynamic cues, the central upregulation in post-infarction RAAS activity serves to initiate this pathway, with Ang-II enhancing TGF-β1 expression in multiple cardiac cell types.

13.3 Where We Are Heading: The Therapeutic Utilization of Identifying Signaling Targets

It is clear that TGF-β is of considerable importance in pathological fibrosis, both in the activation of fibroblasts to a myofibroblastic phenotype, in the induction of ECM gene expression programs, and via the suppression of MMPs [26]. In response, it has been suggested that the inhibition of TGF-β, along with platelet-derived growth factor (PDGF) and several specific cytokines, may have significant therapeutic benefit [28, 29]. In an attempt to block the Ang-II-TGF-β1 pathway, Koh et al. used a synthetic compound (SB 431542) to inhibit activin receptor-like kinase 5 (ALK5), a TGF-β1 receptor [29].

Histopathological assessment and collagen testing were performed to evaluate any potential changes in lung fibrosis. It was concluded that whilst the presence of TGF-β-induced α-SMA was reduced (suggesting decreased fibroblast-to-myofibroblast transition), overall lung fibroblast proliferation exhibited no significant change [29].

With other teams concluding that the TGF-β ligand may not be a useful target, it is clear that inhibitors of ALK5 show at least partial promise [30]. Moreover, with studies subsequent to Koh et al. also identifying a partial blockade of the TGF-β1 pathway via ALK5 targeting, it is clear that further investigations into the effects of inhibiting this receptor are warranted [30].

13.4 Mitochondria and $\Delta\psi_m$

During acute myocardial infarction (AMI), the reduction in perfusion (and thus oxygen) instigates a cascade of deleterious processes that results in cellular apoptosis and necrosis. Particularly affected during this period are mitochondria: double-membrane-bound organelles that catalyze the conversion of chemical energy to ATP via oxidative phosphorylation [31]. Mitochondria are typically well aligned inside CMs, synthesizing high quantities of ATP for energy-consuming proteins such as SERCA and Na^+/K^+ ATPase pumps. In order to continue producing ATP during ischemia, mitochondria switch to lower yield anaerobic respiratory pathways, resulting in the production of lactic acid as a by-product.

The inner mitochondrial membrane potential ($\Delta\psi_m$) refers to the potential difference generated from the transfer of protons across the interior of the mitochondrion's two membranes, usually between −150 mV and −200 mV [31, 32]. $\Delta\psi_m$ reflects integrated mitochondrial capacity and is necessary for the sustenance of oxidative phosphorylation. The inner mitochondrial membrane has a more negative potential difference than the cytoplasm. This means $\Delta\psi_m$ can be effectively measured in vitro using cationic $\Delta\psi_m$-sensitive dyes, with the difference in charge driving the dye into mitochondria, even following washing of the cell preparation [32, 33].

If cardiomyocytes undergo metabolic stress, $\Delta\psi_m$ depolarization begins to occur, uncoupling oxidative phosphorylation in mitochondria. This elicits a switch in the organelle's energy paradigm, whereby instead of producing ATP, they consume it in an attempt to reinstate $\Delta\psi_m$ [32, 33]. The depolarization of $\Delta\psi_m$ in a single mitochondrion triggers the opening of mPTP and IMAC pores on the organelle membrane, both of which are regulated by the mitochondrial benzodiazepine receptor (mBzR) [33]. The opening of these pores allows the release of highly inflammatory reactive oxygen species (ROS) into the local cytoplasmic milieu, propagating $\Delta\psi_m$ depolarization at adjacent mitochondria in a process termed '*ROS-induced ROS release*' [33].

Using high-resolution optical mapping, Akar et al. demonstrated that ischemia and subsequent reperfusion of guinea pig hearts induce electrical destabilization and the induction of ventricular fibrillation, classical of an established phenomenon termed "*ischemia-reperfusion injury*," whereby the post-ischemia restoration of myocardial blood flow induces an upsurge of oxidative stress and arrhythmic activity [32]. However, when repeated with the addition of 4'-chlorodiazepam, an mBzR antagonist, the same reperfusion arrhythmias were not observed.

Mechanistically, the application of an mBzR antagonist prevents the depolarization of $\Delta\psi_m$, resulting in ROS-induced ROS release. This was confirmed by treatment with an mBzR agonist, which exacerbated the reperfusion-induced arrhythmias and directly coupled mitochondrial dysfunction with the acutely pro-arrhythmogenic state of the heart following infarction [32].

Whilst this form of remodeling occurs over a considerably shorter timespan than either the electrophysiological or fibrotic responses to infarction, it is likely of great importance. Indeed, within minutes of ischemia, mitochondrial dysfunction and gap junctional changes occur, both of which are pro-arrhythmic. Clinically, approximately 50% of all deaths in post-MI patients are ascribable to sudden death secondary to sustained VT or VF [34].

13.5   Clinical Implications: Reverse Remodeling of the Mitochondrial Membrane Architecture

Elamipretide (also known as *Bendavia*) is a small tetrapeptide capable of permeating the inner mitochondrial membrane, where it restores and stabilizes cardiolipin: a lipid component of the inner mitochondrial membrane vital for the maintenance of its curved configuration [35]. It has previously been suggested that through this mechanism of action, elamipretide is able to rescue oxidative phosphorylation, preventing or limiting ROS production [36, 37].

The EMBRACE STEMI study [38] compared the efficacy of an early elamipretide prototype to placebo in a multicenter trial of first-time anterior STEMI patients undergoing successful percutaneous coronary intervention (PCI) and found that the drug did not reduce infarct size nor clinical outcome. The veracity of this conclusion is disputable however, with administration of the drug occurring after myocardial revascularization, meaning any $\Delta\psi_m$ depolarization and ROS-induced ROS release that may have been impeded by the compound had already transpired.

In a subsequent study, Daubert et al. evaluated the safety and efficacy of elamipretide in patients with pre-existing heart failure with reduced ejection fraction (HFrEF). No serious adverse effects were observed, with higher doses of the drug significantly reducing LV end-systolic volume (ESV), suggesting favorable effects on cardiac functionality via the restoration of mitochondrial (and thus cellular) energy cycling, in addition to a possible dose–response relationship [39].

┌───┐

Take-Home Message

- Several ionic currents undergo distinct modulation following MI. This can broadly be summarized as a decrease in Na^+/K^+ ATPase and K^+ current magnitude, coupled with an increase in NCX activity.
- Ionic remodeling instigates and maintains aberrant subcellular Ca^{2+} cycling, characterized by defective calcium sequestration into the SR, inefficient release into the cytoplasm, and increase in pro-arrhythmic spark activity.
- Following MI, fibrosis remodeling manifests as a biphasic relationship comprising *replacement* and *reactive* fibrosis. Underlying this longitudinal transition is a nexus of hemodynamic and neurohormonal signaling cues, with the latter using a central Ang-II-TGF-β1 pathway.
- During MI, $\Delta\psi_m$ depolarization uncouples mitochondrial oxidative phosphorylation, switching the organelle to ATP consumers and triggering the opening of membrane mPTP and IMAC pores, allowing the release of reactive oxygen species.

└───┘

References

1. Pons F, Lupón J, Urrutia A, González B, Crespo E, Díez C et al (2010) Mortality and cause of death in patients with heart failure: findings at a specialist multidisciplinary heart failure unit. Rev Española Cardiol (English ed.) 63(3):303–331
2. Pinto JMB, Boyden PA (1999) Electrical remodeling in ischemia and infarction. Cardiovasc Res 42:284
3. Nattel S, Maguy A, Le Bouter S, Yeh Y-H (2007) Arrhythmogenic ion-channel remodeling in the heart: heart failure, myocardial infarction, and atrial fibrillation. Physiol Rev 87:425
4. Zicha S, Xiao L, Stafford S, Cha TJ, Han W, Varro A et al (2004) Transmural expression of transient outward potassium current subunits in normal and failing canine and human hearts. J Physiol 561(3):735–748
5. Klabunde RE (2012) Cardiovascular physiology concepts, 2nd edn. Lippincott Williams & Wilkins, Philadelphia
6. de Brito Santos PE, Barcellos LC, Mill JG, Masuda MO (1995) Ventricular action potential and L-type calcium channel in infarct-induced hypertrophy in rats. J Cardiovasc Electrophysiol 6(11):1004–1014
7. Schmidt U, Hajjar RJ, Helm PA, Kim CS, Doye AA, Gwathmey JK (1998) Contribution of abnormal sarcoplasmic reticulum ATPase activity to systolic and diastolic dysfunction in human heart failure. J Mol Cell Cardiol 30:1929
8. Del Monte F, Harding SE, Schmidt U, Matsui T, Bin KZ, Dec GW et al (1999) Restoration of contractile function in isolated cardiomyocytes from failing human hearts by gene transfer of SERCA2a. Circulation 100(23):2308–2311
9. Li J, Hu SJ, Sun J, Zhu ZH, Zheng X, Wang GZ et al (2005) Construction of phospholamban antisense RNA recombinant adeno-associated virus vector and its effects in rat cardiomyocytes. Acta Pharmacol Sin 26(1):51–55
10. Pogwizd SM, Qi M, Yuan W, Samarel AM, Bers DM (1999) Upregulation of Na+/Ca2+exchanger expression and-function in an arrhythmogenic rabbit model of heart failure. Circ Res 85:1009
11. Wehrens XHT, Lehnart SE, Huang F, Vest JA, Reiken SR, Mohler PJ et al (2003) FKBP12.6 deficiency and defective calcium release channel (ryanodine receptor) function linked to exercise-induced sudden cardiac death. Cell 113:829
12. Schwinger RHG, Bundgaard H, Müller-Ehmsen J, Kjeldsen K (2003) The Na, K-ATPase in the failing human heart. Cardiovasc Res 57:913
13. Stillitano F, Lonardo G, Zicha S, Varro A, Cerbai E, Mugelli A et al (2008) Molecular basis of funny current (If) in normal and failing human heart. J Mol Cell Cardiol 45:289
14. Hoppe UC, Jansen E, Südkamp M, Beuckelmann DJ (1998) Hyperpolarization-activated inward current in ventricular myocytes from normal and failing human hearts. Circulation 97:55
15. Kurata Y, Hisatome I, Matsuda H, Shibamoto T (2005) Dynamical mechanisms of pacemaker generation in IK1- downregulated human ventricular myocytes: insights from bifurcation analyses of a mathematical model. Biophys J 89(4):2865–2887

16. Tomaselli GF, Marbán E (1999) Electrophysiological remodeling in hypertrophy and heart failure. Cardiovasc Res 42:270
17. Fan D, Takawale A, Lee J, Kassiri Z (2012) Cardiac fibroblasts, fibrosis and extracellular matrix remodeling in heart disease. Fibrogenesis Tissue Repair 5:15
18. Shoulders MD, Raines RT (2009) Collagen structure and stability. Annu Rev Biochem 78:929
19. Pauschinger M, Doerner A, Remppis A, Tannhäuser R, Kühl U, Schultheiss HP (1998) Differential myocardial abundance of collagen type I and type III mRNA in dilated cardiomyopathy: effects of myocardial inflammation. Cardiovasc Res 37:123
20. Mouw JK, Ou G, Weaver VM (2014) Extracellular matrix assembly: a multiscale deconstruction. Nat Rev Mol Cell Biol 15:771
21. Baum J, Duffy HS (2011) Fibroblasts and myofibroblasts: what are we talking about? J Cardiovasc Pharmacol 57:376
22. Nagase H, Visse R, Murphy G (2006) Structure and function of matrix metalloproteinases and TIMPs. Cardiovasc Res 69:562
23. Yue L, Xie J, Nattel S (2011) Molecular determinants of cardiac fibroblast electrical function and therapeutic implications for atrial fibrillation. Cardiovasc Res 89:744
24. Rohr S (2012) Arrhythmogenic implications of fibroblast-myocyte interactions. Circ Arrhythmia Electrophysiol 5:442
25. Nguyen TP, Qu Z, Weiss JN (2014) Cardiac fibrosis and arrhythmogenesis: the road to repair is paved with perils. J Mol Cell Cardiol 70:83
26. Meng XM, Nikolic-Paterson DJ, Lan HY (2016) TGF-β: the master regulator of fibrosis. Nat Rev Nephrol 12:325–338
27. Chemaly ER, Kang S, Zhang S, Mccollum L, Chen J, Bénard L et al (2013) Differential patterns of replacement and reactive fibrosis in pressure and volume overload are related to the propensity for ischaemia and involve resistin. J Physiol 591(21):5337–5355
28. Tojo M, Hamashima Y, Hanyu A, Kajimoto T, Saitoh M, Miyazono K et al (2005) The ALK-5 inhibitor A-83-01 inhibits Smad signaling and epithelial-to-mesenchymal transition by transforming growth factor-β. Cancer Sci 96:791
29. Koh RY, Lim CL, Uhal BD, Abdullah M, Vidyadaran S, Ho CC et al (2015) Inhibition of transforming growth factor-β via the activin receptor-like kinase-5 inhibitor attenuates pulmonary fibrosis. Mol Med Rep 11:3808
30. Nagaraj NS, Datta PK (2010) Targeting the transforming growth factor-β signaling pathway in human cancer. Expert Opin Investig Drugs 19:77
31. Sack MN (2006) Mitochondrial depolarization and the role of uncoupling proteins in ischemia tolerance. Cardiovasc Res 72:210
32. Akar FG, Aon MA, Tomaselli GF, O'Rourke B (2005) The mitochondrial origin of postischemic arrhythmias. J Clin Invest 115:3527
33. Lyon AR, Joudrey PJ, Jin D, Nass RD, Aon MA, O'Rourke B et al (2010) Optical imaging of mitochondrial function uncovers actively propagating waves of mitochondrial membrane potential collapse across intact heart. J Mol Cell Cardiol 49:565
34. Bhar-Amato J, Davies W, Agarwal S (2017) Ventricular arrhythmia after acute myocardial infarction: 'the perfect storm'. Arrhythmia Electrophysiol Rev 6:134
35. Karaa A, Haas R, Goldstein A, Vockley J, Weaver WD, Cohen BH (2018) Randomized dose-escalation trial of elamipretide in adults with primary mitochondrial myopathy. Neurology 90:e1212
36. Kloner RA, Hale SL, Dai W, Gorman RC, Shuto T, Koomalsingh KJ et al (2012) Reduction of ischemia/reperfusion injury with bendavia, a mitochondria-targeting cytoprotective peptide. J Am Heart Assoc 1:e001644
37. Szeto HH (2014) First-in-class cardiolipin-protective compound as a therapeutic agent to restore mitochondrial bioenergetics. Br J Pharmacol 171:2029
38. Gibson CM, Giugliano RP, Kloner RA, Bode C, Tendera M, Jánosi A et al (2016) EMBRACE STEMI study: a phase 2a trial to evaluate the safety, tolerability, and efficacy of intravenous MTP-131 on reperfusion injury in patients undergoing primary percutaneous coronary intervention. Eur Heart J 37:1296
39. Daubert MA, Yow E, Dunn G, Marchev S, Barnhart H, Douglas PS et al (2017) Novel mitochondria-targeting peptide in heart failure treatment: a randomized, placebo-controlled trial of Elamipretide. Circ Hear Fail 10:e004389

13

Developments in Heart Failure: Mechanical Unloading with LVADs, Exosomes, and MicroRNAs

Samuel Guymer and Mayooran Shanmuganathan

© Springer Nature Switzerland AG 2019
C. Terracciano, S. Guymer (eds.), *Heart of the Matter*, Learning Materials in Biosciences,
https://doi.org/10.1007/978-3-030-24219-0_14

What You Will Learn in This Chapter

This chapter will begin by briefly establishing the global clinical burden of heart failure (HF) and limitations of current treatment regimens. The use of left ventricular assist devices (LVADs) in advanced HF will then be explored, including a discussion of the underlying principles of mechanical unloading. We will then evaluate the emerging evidence of the role that exosomes and microRNAs may have in the diagnosis and treatment of cardiovascular diseases, including myocardial infarction.

Learning Objectives
- Provide an overview of current HF therapeutics and describe the unmet needs.
- Describe the beneficial effects of LVAD therapy and how newer models address some of their inherent deficiencies.
- Gain insights into the emerging field of research on exosomes and microRNAs.

14.1 Heart Failure: The Need for New Thinking

The European Society of Cardiology (2016) defines heart failure (HF) as "a clinical syndrome characterised by typical symptoms that may be accompanied by signs caused by a structural and/or functional cardiac abnormality, resulting in a reduced cardiac output (CO) and/or elevated intracardiac pressure at rest or during stress" [1]. Fundamentally, HF manifests as the convergent final pathway of a broad potpourri of cardiovascular (CV) pathologies including hypertension and myocardial infarction (MI), the two most common etiologies of this syndrome, in addition to arrhythmias, cardiomyopathies, and others [2].

As previously discussed, these serve as "insults" to the myocardium that precipitates a reduction in the quantity and/or functional capacity of cardiomyocytes, triggering a multiplex remodeling apparatus mediated in part by chronic β-adrenergic activation and the renin–angiotensin–aldosterone system (RAAS).

With individuals aged 65 years and older comprising ~80% of HF cases, the clinical burden presented by this syndrome is only expected to increase when taken in the context of an aging global population [3]. Furthermore, the American Heart Association (AHA) predicts that between 2012 and 2030, the number of cases in the United States will increase by an estimated 46%, with an associated 127% increase in healthcare costs, hinting at both a rapidly expanding patient base and escalating fiscal burden [4]. The current therapeutic paradigm is aimed at (1) alleviating the symptoms of HF, (2) delaying its progression, and ultimately (3) reducing mortality [1]. In an attempt to achieve this, both pharmacological and device therapies are employed.

Current guidelines prescribe the use of diuretics in all HF patients with signs of congestion, irrespective of their LV ejection fraction (LVEF) [1]. In HF with reduced LVEF (HFrEF) patients, the mainstay of pharmacological therapy includes an angiotensin-converting enzyme (ACE) inhibitor (or angiotensin-II receptor antagonist or angiotensin receptor-neprilysin inhibitor/ARNI), a beta (β)-blocker, and an aldosterone antagonist, all of which improve prognosis [1]. Ivabradine, digoxin, hydralazine, and isosorbide dinitrate are also used in some instances. The pharmacological basis of some of these therapies will be discussed further in ▶ Chap. 15.

Although less frequently used, device therapies constitute a vital dimension of treatment. As we previously discussed, the CARE-HF study recruited symptomatic HFrEF

14

patients with a wide QRS to receive either cardiac resynchronization therapy – pacemaker (CRT with a pacemaker/CRT-P) or not, with the CRT-P group demonstrating significant reductions in both all-cause mortality and unplanned hospitalizations for major CV events [5].

Registry data suggest that survival on optimal medical therapy is ~50% at 5 years following the diagnosis of HF [6]. Thus, when medications and pacemaker device interventions fail to control the symptoms of HF, patients should be referred to specialist centers to be considered for ventricular assist device (VAD) therapies or heart transplantation [1].

14.2 Ventricular Assist Devices

Ventricular assist devices (VADs) are artificial appliances that assist either the left (LVAD) or both (BiVAD) failing ventricles by partially or completely replacing cardiac function [7]. With cardiac explantation not required, they are distinct from total artificial hearts (TAH). In advanced HF patients requiring mechanical support, VADs form one branch of therapy, along with TAH and heart transplantation. From implantation of the earliest VAD by Dr. Michael E. DeBakey in 1966, technological advancement has facilitated a progressive refinement of the LVAD, with newer models being smaller (and thus more mobile) than their progenitors, having higher capacity, and being more durable [7].

Mechanistically, LVADs are implanted at the apex of the left ventricle to accept atrial blood via an inflow cannula into a motorized impeller. In most models, the impeller acts as a centrifugal pump to accelerate the outward movement of fluid, propelling it at high velocity through an outflow graft into the aorta [7, 8]. Different models have varying propulsion mechanisms, with some employing an axial flow system, while almost all of the newer generation VADs act as continuous-flow pumps, meaning no pulse is palpable in the patient. Finally, a percutaneous driveline connects the device to an external system controller that drives impeller movement and monitors pump function. This design is summarized in ◘ Fig. 14.1 [9]. Two widely used VADs are the HeartWare and the HeartMate II [9, 11].

By assuming the pumping function of the failing ventricles, VADs are able to restore cardiac output and organ perfusion, in addition to reducing mechanical stress [11]. Moreover, in a small proportion of patients, this promotes "reverse remodeling" of the functionally overloaded and remodeled heart, visible morphologically as reduced myocardial hypertrophy [12, 13]. Indeed, Terracciano et al. observed this phenomenon in cardiomyocytes extracted from tissue biopsies of LVAD patients; however, they argued that this was in fact precipitated by changes in excitation–contraction coupling rather than reduction in workload directly, with decreased cardiomyocyte (CM) hypertrophy having a lesser impact on recovery [14].

This proposal is broadly congruent with other studies, which similarly identify a weak association between hypertrophic CM changes in HF and alterations in functional capacity [15]. This finding, amongst others, has given rise to a considerable volume of research on the effects of mechanical unloading in the failing heart, including studies in humans, animal models, and individual CMs [16, 17]. Indeed, chronic unloading (i.e., the mechanism of LVADs) has been shown to induce significant alterations in subcellular Ca^{2+} cycling [18]. In a seminal paper by Ibrahim et al., chronic unloading facilitated a redistribution of T-tubules in rat LV CMs, structurally uncoupling L-type Ca^{2+} channels from

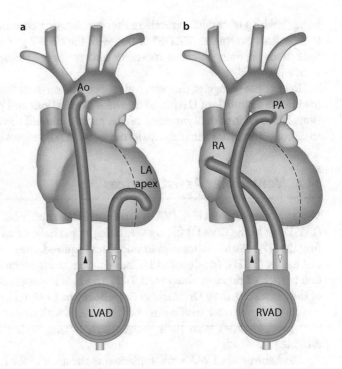

Fig. 14.1 Schematic illustration of the placement of inflow and outflow cannulae for **a** LVADs and **b** RVADs. Abbreviations: Ao aorta, LV left ventricle, LVAD left ventricular assist device, PA pulmonary artery, RA right atrium, RVAD right ventricular assist device. (Permission to reuse from the study by Krishnamani et al. [9]. Image from [10])

ryanodine receptors in order to increase spark activity and Ca^{2+} transient asynchronicity [18]. Human studies have yielded a range of mechanistic conclusions, with the disparity likely due to factors such as the duration of unloading, type, and cause of HF, in addition to the incidence of complications not related to unloading [17, 19].

LVAD-related complications cause major morbidities [20]. In a study of patients who received HeartMate II LVADs as destination therapy and followed up 2 years later, adverse events included bleeding (54% of patients), driveline infection (~19%), thromboembolic events (~12%), RV heart failure (~18%), and mechanical failure requiring replacement (4% patients) [21]. This high rate of adverse events further limits the clinical utilization of mechanical unloading. Overall, however, three conclusions can be drawn from worldwide experience of using LVADs in advanced HF patients:

1. LVAD therapy in advanced HF (acute or chronic) achieves superior outcomes compared to standard medical therapy [8].
2. LVAD therapy in advanced chronic HF is unlikely to improve LV function.
3. LVAD therapy in acute or recent-onset advanced HF may achieve improvement in LV function. This is primarily because a high proportion of recent-onset HF that rapidly deteriorates is caused by reversible pathologies such as infective, inflammatory, or postpartum cardiomyopathy, for which mechanical unloading is effective [22]. During the period of unloading, the underlying insult may no longer be present (e.g., following an immune response), allowing the LVAD to reduce the cardiac workload whilst maintaining the perfusion of other organs [22].

Subsequently, transplantation remains the gold standard intervention for end-stage HF, with the role of VADs being as a "bridge to transplantation," sustaining organ (especially lung and renal) function until transplantation can occur [23, 24]. However, in the context of a global deficit of donor hearts, as eluded to above, newer VADs have been proposed as a potential destination therapy [25], with engineering advances catalyzing the development of newer devices that seek to address existing limitations [7].

One such example is the HeartMate 3, which received regulatory approval in 2018 and features a novel programming algorithm that creates dynamic pressure patterns more representative of the heart [10, 26]. The salient feature of HeartMate 3 is a fully magnetically levitated ("MagLev") centrifugal-flow rotor, which combats a central problem of current continuous-flow LVADs, namely the high shear forces generated by the pumping apparatus [26]. By minimizing the shearing of blood, the MagLev rotor is reported to reduce hemolysis and the cleavage of von Willebrand factor, diminishing the probability of thrombosis forming around the rotor and the subsequent implications this may have with regard to strokes and bleeding events [26].

In the MOMENTUM 3 trial, 152 patients were assigned to the HeartMate 3 LVAD (magnetically levitated continuous-flow centrifugal pump) and 142 to the HeartMate II LVAD (axial-flow continuous-flow pump), all of whom had advanced heart failure [10]. Implantation of the HeartMate 3 was associated with a reduced risk of stroke, or reoperation to replace or remove the device at 6 months. While no significant difference was observed in GI bleeding rates or survival, it was notable that no significant suspected nor confirmed pump thrombosis was identified in the HeartMate 3 group, suggesting that the main advantage of a magnetically levitated centrifugal rotor is life quality improvement rather than prolonged survival [10].

The MOMENTUM 3 trialists published a 2-year follow-up study in 2018 showing a survival rate of 83% and sustained improvement in the HeartMate 3 group in the incidence of stroke (10%) and pump thrombosis (1%) when compared to HeartMate 2 [27].

14.3 Emerging Research in HF and Ischemic Heart Disease – Exosomes and MicroRNAs

We have previously discussed the multifaceted remodeling process that characterizes the development of HF. As the signaling underlying these pathways is increasingly elucidated, interest is converging on the targeting of specific molecules in order to arrest or even reverse pathogenic remodeling. Of particular interest are exosomes: a nano-sized (30–100 nm) subset of extracellular vesicle (EV) secreted by myriad cell types and associated with intercellular communication via the delivery of cargo including nucleic acids, lipids, and microRNAs [28, 29]. There is emerging evidence on the role that exosomes play in several cardiac conditions, including ischemic heart disease [29, 30].

Exosome biogenesis is characterized by the binding and invagination of cargo through the cellular plasma membrane, after which it is transported to early endosomes where a similar inward budding process occurs at the endosomal delimiting membrane [30]. This loading process permits a wide range of potential cargo to enter the exosome and is mediated by a plethora of cellular machinery, with the endosomal sorting complex required for transport (ESCRT) of particular importance [31]. Unless destined for degradation, the resulting multivesicular bodies (MVBs) constitutively fuse with the plasma membrane in a process orchestrated by Rab GTPases, resulting in the pulsatile exocytosis of their intraluminal vesicle (ILV) cargo into the extracellular milieu [32, 33].

MicroRNAs (miRNAs/miRs) are one such type of this cargo: short (17–25 nucleotides), non-coding RNA sequences that function as post-transcriptional regulators of gene expression through the induction of mRNA degradation or repression of translation [34]. The regulatory role of miRNAs is evidenced by extensive gene targeting, in addition to evolutionary conservation, with the aberrant expression of several miRNAs associated with pathological states, especially the progression of cancer [35, 36]. Broadly speaking, miRNAs are synthesized as polynucleotide duplexes of 5′ and 3′ strands, of which one has enhanced bioactivity and is therefore referred to as the guide strand. Following removal of the non-guide (passenger) strand, the remaining guide strand incorporates with Argonaute proteins to form an RNA-induced silencing complex (RISC) that ultimately mediates the silencing of mRNA expression [37].

14.4 Exosomes as a Therapeutic Delivery Apparatus

Through the transport of miRs and other signaling cargo, exosomes have an integral role as an endogenous mediator of intercellular communication, both at a paracrine level via their direct release into the local interstitium and in an endocrine capacity through plasma secretion [38]. There is emerging strong evidence that exosomes (shown in ◘ Fig. 14.2) secreted from various types of stem cells induce cardioprotection [40–42]. Studies have also demonstrated the functional importance of endogenous cardiac exosomal crosstalk between different cardiac cell types in mammalian hearts [43], and that this may be modified by comorbidities such as diabetes mellitus [44].

Research on stem cell exosomes is providing evidence that they can recapitulate the effects of their parent cells, thus avoiding the need for cell therapy and the associated disadvantages. Exosomes are robust and their contents are well protected by a lipid bilayer membrane; they can be stored "on-the-shelf" at −80 °C for prolonged periods and

14

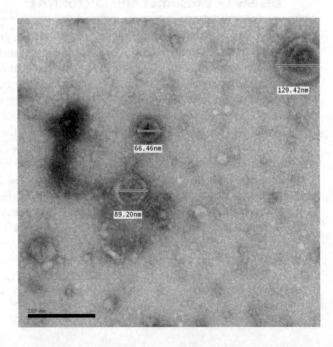

◘ **Fig. 14.2** Exosomes isolated from a freshly explanted murine heart in our lab. Based on a published method [39], fresh tissue was left in six-well tissue culture plates with M199-based medium and incubated in 5% CO_2 at 37 °C for 24 hours. Exosomes were isolated from the conditioned medium using serial ultracentrifugation and viewed under transmission electron microscope (TEM) using standard protocols. Note that the exosomes possess a typical cup-shaped lipid bilayer membrane. (Credit for TEM: Dr Padmini Sarathchandra, Magdi Yacoub Heart Centre, Harefield, UK.)

bioengineered to contain a desired array of intracellular cargo [39]. Moreover, via the membrane expression of a specific array of ligands, exosomes have exhibited cell-specific delivery of their cargo without the need of additional vectors [45]. Although exosomes carry MHC molecules on their surface, their immunogenicity is considered to be low [45], allowing the possibility of repeated allogenic doses.

Thus, it is attractive to design bioengineered exosomes as a novel nanotransport apparatus for the targeted delivery of drugs and miRs [30, 46, 47]. This might overcome the problems encountered in previously used tools such as liposomes [48, 49]. However, clinical application of the exosomal delivery of nucleic acids and other therapeutics is yet to be realized [30].

14.5 MicroRNAs as a Biomarker of Myocardial Infarction

A biomarker is any endogenous molecule, gene, or feature that can be utilized to identify the presence or magnitude of a specific physiological or pathological entity. In cardiovascular medicine, biomarkers have proved a critical diagnostic apparatus, perhaps best exemplified in myocardial infarction (MI) – one of the foremost HF etiologies. Currently, diagnosis relies on the elevation of the biomarker cardiac troponin (cTn) in the plasma, which is released by dying cardiomyocytes and detectably elevated 4–6 hours later [50, 51].

Fast confirmation is critical in MI, with diagnostic expediency facilitating the rapid initiation of reperfusion interventions. These are recommended to commence within 90 minutes of hospital arrival for maximum prognostic benefit [51]. The delayed plasma elevation of cTn has resulted in therapeutic guidelines indicating the commencement of reperfusion prior to the confirmation of diagnosis. Despite the development of point-of-care higher sensitivity troponin assays, there remains a need for novel biomarkers of MI that exhibit faster and reliable detectability.

The discovery that exosomal cargo is not static, instead reflecting the physiological or pathological state of their origin cell over time, has focused interest on changes in exosomal cargo during myocardial ischemia and infarction [32]. One cargo fraction known to undergo significant expression changes in a variety of pathologies is miRNAs, which are released by the myocardium into local tissue and the circulation [36]. In the plasma, miR-NAs circulate either independently, where they are prone to degradation by RNases, in complexes with Argonaute proteins, where they are conferred a relative protection, or in exosomes [37, 52].

14.5.1 The Example of MicroRNA-133a

In one study to investigate the feasibility of miRNAs as biomarkers of infarction, plasma was acquired from patients with acute ST-segment elevation MI (STEMI) during the onset of symptoms and following reperfusion, with the expression of several previously identified miRNA candidates measured [53]. The plasma expressions of miR-1, -133a, -133b, and -499-5p were found to be significantly upregulated after the onset of MI symptoms, decaying back to baseline within the following 5 days. In the same study, MI was then induced in a mouse model via coronary ligation. The same four miRs exhibited similar expression changes over time, namely an increase following the induction of MI and a gradual decay thereafter [53].

Further to this, Kuwabara et al. investigated human and mouse miRNA changes post-injury, recruiting 71 human patients with a variety of CV pathologies, not just ST-elevation MI [54]. The serum levels of miR-1 and -133a were dramatically upregulated in acute MI, with the latter also in patients with unstable angina pectoris and Takotsubo cardiomyopathy, without any elevation of serum creatine phosphokinase or cTn [54]. This reaffirms the notion that unlike cardiac troponin, the release of miR-133a is mediated by cardiac injury potentially linked to ischemia, and not by necrosis alone.

This would suggest that during infarction, miR-133a is secreted into plasma by the ischemic myocardium as an early reactive response to injury, prior to the onset of necrosis associated with cTn release. As such, this is congruent with the elevation of plasma miR-1 and -133a noted shortly (<2 hours) after the onset of chest pain in MI and significantly before any elevation in cTn was observed (>6 hours) [54]. Use of miR-133a as a biomarker of "pre-necrosis MI" in humans is further supported by D'Alessandra et al., who similarly noted this phenomenon [53].

However, given the unpredictable degradation dynamics of non-complex-bound miR-133a, its use as a biomarker would pivot on its exosomal release by the injured myocardium. To investigate this, Kuwabara et al. used the Ca^{2+} ionophore A23187 to stimulate H9c2 cardiomyoblasts in vitro, mimicking the upsurge in cytoplasmic Ca^{2+} observed in pre-necrosis myocardial injury. Exosomes were isolated from the resulting culture medium and screened for miR-133a, where a dose-dependent release relationship was observed. Principally, this identifies cytoplasmic signaling typical of myocardial injury as a trigger for the exosomal release of miR-133a, a finding supported by murine studies associating STEMI-induced myocardial injury with the elevated release of extracellular vesicles [55, 56].

Of course, this still fails to completely address the issue of free circulating miR-133a in the plasma, which may have the potential to distort the plasma expression data of miR-133a sufficiently to limit its use as a biomarker. Several studies have now demonstrated the sensitivity of circulating miRNAs as biomarkers of myocardial injury, particularly miR-133a. However, with many miRNA signaling targets yet to be elucidated, clearly further research is warranted. Regardless, exosomal miRs may yet hold much promise as novel biomarkers of pre-infarction clinical states, as well as in established MI.

14

Take-Home Message

- HF represents an escalating global healthcare burden, with current treatments targeting the management of symptoms and delaying progression.
- LVAD therapy in advanced HF produces a spectrum of clinical outcome trajectories, resulting in its use as a "bridge to transplantation," a "bridge to recovery," or in some instances as a "bridge to destination."
- Exosomes are naturally occurring extracellular vesicles that appear to exert paracrine effects by shuttling various cargoes between cells, one of which is microRNA. Although research is in its infancy, they offer promise as diagnostic tools and drug delivery vehicles.
- MicroRNAs appear to play a significant role in many bodily processes, including MI and HF. miR-133a is secreted by the ischemic myocardium prior to the onset of infarction, suggesting its use as a biomarker of "pre-necrosis" MI.

References

1. Ponikowski P, Voors AA, Anker SD, Bueno H, Cleland JGF, Coats AJS et al (2016) 2016 ESC guidelines for the diagnosis and treatment of acute and chronic heart failure. Eur Heart J 37(27):2129–2200. https://doi.org/10.1093/eurheartj/ehw128
2. Inamdar AA, Inamdar AC (2016) Heart failure: diagnosis, management and utilization. J Clin Med 5(7):62. https://doi.org/10.3390/jcm5070062
3. Masoudi FA, Havranek EP, Krumholz HM (2002) The burden of chronic congestive heart failure in older persons: magnitude and implications for policy and research. Heart Fail Rev 7(1):9–16. https://doi.org/10.1023/A:1013793621248
4. Heidenreich PA, Albert NM, Allen LA, Bluemke DA, Butler J, Fonarow GC et al (2013) Forecasting the impact of heart failure in the United States: a policy statement from the American Heart Association. Circ Heart Fail 6(3):606–619. https://doi.org/10.1161/HHF.0b013e318291329a
5. Cleland JG, Daubert JC, Erdmann E, Freemantle N, Gras D, Kappenberger L et al (2005) The effect of cardiac resynchronization on morbidity and mortality in heart failure. N Engl J Med 352(15):1539–1549. https://doi.org/10.1056/NEJMoa050496
6. Zarrinkoub R, Wettermark B, Wändell P, Mejhert M, Szulkin R, Ljunggren G et al (2013) The epidemiology of heart failure, based on data for 2.1 million inhabitants in Sweden. Eur J Heart Fail 15(9):995–1002. https://doi.org/10.1093/eurjhf/hft064
7. Samak M, Fatullayev J, Sabashnikov A, Zeriouh M, Rahmanian PB, Choi YH et al (2015) Past and present of total artificial heart therapy: a success story. Med Sci Monit Basic Res 21:183–190. https://doi.org/10.12659/MSMBR.895418
8. Rose EA, Gelijns AC, Moskowitz AJ, Heitjan DF, Stevenson LW, Dembitsky W et al (2001) Long-term use of a left ventricular assist device for end-stage heart failure. N Engl J Med 345(20):1435–1443. https://doi.org/10.1056/NEJMoa012175
9. Krishnamani R, Denofrio D, Konstam MA (2010) Emerging ventricular assist devices for long-term cardiac support. Nat Rev Cardiol 7(2):71–76. https://doi.org/10.1038/nrcardio.2009.222
10. Mehra MR, Naka Y, Uriel N, Goldstein DJ, Cleveland JC, Colombo PC et al (2017) A fully magnetically levitated circulatory pump for advanced heart failure. N Engl J Med 376(5):440–450. https://doi.org/10.1056/NEJMoa1610426
11. Prinzing A, Herold U, Berkefeld A, Krane M, Lange R, Voss B (2016) Left ventricular assist devices-current state and perspectives. J Thorac Dis 8(8):E660–E666. https://doi.org/10.21037/jtd.2016.07.13
12. Simon MA, Kormos RL, Murali S, Nair P, Heffernan M, Gorcsan J et al (2005) Myocardial recovery using ventricular assist devices: prevalence, clinical characteristics and outcomes. Circulation 112(9 Suppl):I32–I36. https://doi.org/10.1161/CIRCULATIONAHA.104.524124.
13. Birks EJ, Tansley PD, Hardy J, George RS, Bowles CT, Burke M et al (2006) Left ventricular assist device and drug therapy for the reversal of heart failure. N Engl J Med 355(18):1873–1884. https://doi.org/10.1056/NEJMoa053063
14. Terracciano CM, Hardy J, Birks EJ, Khaghani A, Banner NR, Yacoub MH (2004) Clinical recovery from end-stage heart failure using left-ventricular assist device and pharmacological therapy correlates with increased sarcoplasmic reticulum calcium content but not with regression of cellular hypertrophy. Circulation 109(19):2263–2265. https://doi.org/10.1161/01.CIR.0000129233.51320.92
15. Botchway AN, Turner MA, Sheridan DJ, Flores NA, Fry CH (2003) Electrophysiological effects accompanying regression of left ventricular hypertrophy. Cardiovasc Res 60(3):510–517. https://doi.org/10.1016/j.cardiores.2003.08.013
16. Gustafsson F, Rogers JG (2017) Left ventricular assist device therapy in advanced heart failure: patient selection and outcomes. Eur J Heart Fail 19(5):595–602. https://doi.org/10.1002/ejhf.779
17. Drakos SG, Terrovitis JV, Anastasiou-Nana MI, Nanas JN (2007) Reverse remodeling during long-term mechanical unloading of the left ventricle. J Mol Cell Cardiol 43(3):231–242. https://doi.org/10.1016/j.yjmcc.2007.05.020
18. Ibrahim M, Al Masri A, Navaratnarajah M, Siedlecka U, Soppa GK, Moshkov A et al (2010) Prolonged mechanical unloading affects cardiomyocyte excitation-contraction coupling, transverse-tubule structure, and the cell surface. FASEB J 24(9):3321–3329. https://doi.org/10.1096/fj.10-156638
19. Rossing K, Gustafsson F (2018) Medical and mechanical unloading in advanced heart failure: hope for cardiac recovery? Eur J Heart Fail 20(1):175–177. https://doi.org/10.1002/ejhf.1018
20. Wever-Pinzon O, Drakos SG, Kfoury AG, Nativi JN, Gilbert EM, Everitt M et al (2013) Morbidity and mortality in heart transplant candidates supported with mechanical circulatory support: is reap-

praisal of the current United network for organ sharing thoracic organ allocation policy justified? Circulation 127(4):452–462. https://doi.org/10.1161/CIRCULATIONAHA.112.100123

21. Jorde UP, Kushwaha SS, Tatooles AJ, Naka Y, Bhat G, Long JW et al (2014) Results of the destination therapy post-food and drug administration approval study with a continuous flow left ventricular assist device: a prospective study using the INTERMACS registry (Interagency Registry for Mechanically Assisted Circulatory Support). J Am Coll Cardiol 63(17):1751–1757. https://doi.org/10.1016/j.jacc.2014.01.053

22. Loyaga-Rendon RY, Pamboukian SV, Tallaj JA, Acharya D, Cantor R, Starling RC et al (2014) Outcomes of patients with peripartum cardiomyopathy who received mechanical circulatory support. Data from the Interagency Registry for Mechanically Assisted Circulatory Support. Circ Hear Fail 7(2):300–309. https://doi.org/10.1161/CIRCHEARTFAILURE.113.000721

23. Iwashima Y, Yanase M, Horio T, Seguchi O, Murata Y, Fujita T et al (2012) Effect of pulsatile left ventricular assist system implantation on Doppler measurements of renal hemodynamics in patients with advanced heart failure. Artif Organs 36(4):353–358. https://doi.org/10.1111/j.1525-1594.2011.01351.x

24. John R, Pagani FD, Naka Y, Boyle A, Conte JV, Russell SD et al (2010) Post-cardiac transplant survival after support with a continuous-flow left ventricular assist device: impact of duration of left ventricular assist device support and other variables. J Thorac Cardiovasc Surg 140(1):174–181. https://doi.org/10.1016/j.jtcvs.2010.03.037

25. Mancini D, Colombo PC (2015) Left ventricular assist devices: a rapidly evolving alternative to transplant. J Am Coll Cardiol 65(23):2542–2555. https://doi.org/10.1016/j.jacc.2015.04.039

26. Heatley G, Sood P, Goldstein D, Uriel N, Cleveland J, Middlebrook D et al (2016) Clinical trial design and rationale of the Multicenter Study of MagLev Technology in Patients Undergoing Mechanical Circulatory Support Therapy with HeartMate 3 (MOMENTUM 3) investigational device exemption clinical study protocol. J Hear Lung Transplant 35(4):528–536. https://doi.org/10.1016/j.healun.2016.01.021

27. Mehra MR, Goldstein DJ, Uriel N, Cleveland JC, Yuzefpolskaya M, Salerno C et al (2018) Two-year outcomes with a magnetically levitated cardiac pump in heart failure. N Engl J Med 378(15):1386–1395. https://doi.org/10.1056/NEJMoa1800866

28. Valadi H, Ekström K, Bossios A, Sjöstrand M, Lee JJ, Lötvall JO (2007) Exosome-mediated transfer of mRNAs and microRNAs is a novel mechanism of genetic exchange between cells. Nat Cell Biol 9(6):654–659. https://doi.org/10.1038/ncb1596

29. Kumarswamy R, Thum T (2013) Non-coding RNAs in cardiac remodeling and heart failure. Circ Res 113(6):676–689. https://doi.org/10.1161/CIRCRESAHA.113.300226

30. Shanmuganathan M, Vughs J, Noseda M, Emanueli C (2018) Exosomes: basic biology and technological advancements suggesting their potential as ischemic heart disease therapeutics. Front Physiol 9:1159. https://doi.org/10.3389/fphys.2018.01159

31. Colombo M, Moita C, van Niel G, Kowal J, Vigneron J, Benaroch P et al (2013) Analysis of ESCRT functions in exosome biogenesis, composition and secretion highlights the heterogeneity of extracellular vesicles. J Cell Sci 126(24):5553–5565. https://doi.org/10.1242/jcs.128868

32. Edgar JR, Eden ER, Futter CE (2014) Hrs- and CD63-dependent competing mechanisms make different sized endosomal intraluminal vesicles. Traffic 15(2):197–211. https://doi.org/10.1111/tra.12139

33. Trajkovic K, Hsu C, Chiantia S, Rajendran L, Wenzel D, Wieland F et al (2008) Ceramide triggers budding of exosome vesicles into multivesicular endosomes. Science 319(5867):1244–1247. https://doi.org/10.1126/science.1153124

34. Han J, Lee Y, Yeom KH, Nam JW, Heo I, Rhee JK et al (2006) Molecular basis for the recognition of primary microRNAs by the Drosha-DGCR8 complex. Cell 125(5):887–901. https://doi.org/10.1016/j.cell.2006.03.043

35. Takane K, Fujishima K, Watanabe Y, Sato A, Saito N, Tomita M et al (2010) Computational prediction and experimental validation of evolutionarily conserved microRNA target genes in bilaterian animals. BMC Genomics 11:101. https://doi.org/10.1186/1471-2164-11-101

36. Singh R, Pochampally R, Watabe K, Lu Z, Mo YY (2014) Exosome-mediated transfer of miR-10b promotes cell invasion in breast cancer. Mol Cancer 13:256. https://doi.org/10.1186/1476-4598-13-256

37. Chendrimada TP, Gregory RI, Kumaraswamy E, Norman J, Cooch N, Nishikura K et al (2005) TRBP recruits the Dicer complex to Ago2 for microRNA processing and gene silencing. Nature 436(7051):740–744. https://doi.org/10.1038/nature03868

14

38. Kim H, Yun N, Mun D, Kang JY, Lee SH, Park H et al (2018) Cardiac-specific delivery by cardiac tissue-targeting peptide-expressing exosomes. Biochem Biophys Res Commun 499(4):803–808. https://doi.org/10.1016/j.bbrc.2018.03.227

39. Mincheva-Nilsson L, Baranov V, Nagaeva O, Dehlin E (2016) Isolation and characterization of exosomes from cultures of tissue explants and cell lines. Curr Protoc Immunol 115:14.42.1–14.42.21. https://doi.org/10.1002/cpim.17

40. Barile L, Moccetti T, Marbán E, Vassalli G (2017) Roles of exosomes in cardioprotection. Eur Heart J 38(18):1372–1379. https://doi.org/10.1093/eurheartj/ehw304

41. Eulalio A, Mano M, Dal Ferro M, Zentilin L, Sinagra G, Zacchigna S et al (2012) Functional screening identifies miRNAs inducing cardiac regeneration. Nature 492(7429):376–381. https://doi.org/10.1038/nature11739

42. Khan M, Nickoloff E, Abramova T, Johnson J, Verma SK, Krishnamurthy P et al (2015) Embryonic stem cell-derived exosomes promote endogenous repair mechanisms and enhance cardiac function following myocardial infarction. Circ Res 117(1):52–64. https://doi.org/10.1161/CIRCRESAHA.117.305990

43. Garcia NA, Moncayo-Arlandi J, Sepulveda P, Diez-Juan A (2016) Cardiomyocyte exosomes regulate glycolytic flux in endothelium by direct transfer of GLUT transporters and glycolytic enzymes. Cardiovasc Res 109(3):397–408. https://doi.org/10.1093/cvr/cvv260

44. Wang X, Huang W, Liu G, Cai W, Millard RW, Wang Y et al (2014) Cardiomyocytes mediate anti-angiogenesis in type 2 diabetic rats through the exosomal transfer of miR-320 into endothelial cells. J Mol Cell Cardiol 74:139–150. https://doi.org/10.1016/j.yjmcc.2014.05.001

45. Alvarez-Erviti L, Seow Y, Yin H, Betts C, Lakhal S, Wood MJ (2011) Delivery of siRNA to the mouse brain by systemic injection of targeted exosomes. Nat Biotechnol 29(4):341–345. https://doi.org/10.1038/nbt.1807

46. Kishore R, Khan M (2016) More than tiny sacks: stem cell exosomes as cell-free modality for cardiac repair. Circ Res 118(2):330–343. https://doi.org/10.1161/CIRCRESAHA.115.307654

47. Marbán E (2018) The secret life of exosomes: what bees can teach us about next-generation therapeutics. J Am Coll Cardiol 71(2):193–200. https://doi.org/10.1016/j.jacc.2017.11.013

48. Stremersch S, Vandenbroucke RE, Van Wonterghem E, Hendrix A, De Smedt SC, Raemdonck K (2016) Comparing exosome-like vesicles with liposomes for the functional cellular delivery of small RNAs. J Control Release 232:51–61. https://doi.org/10.1016/j.jconrel.2016.04.005

49. Yoshida K, Burton GF, McKinney JS, Young H, Ellis EF (1992) Brain and tissue distribution of polyethylene glycol-conjugated superoxide dismutase in rats. Stroke 23(6):865–869. https://doi.org/10.1161/01.STR.23.6.865

50. Babuin L, Jaffe AS (2005) Troponin: the biomarker of choice for the detection of cardiac injury. CMAJ 173(10):1191–1202. https://doi.org/10.1503/cmaj.050141

51. Stirrup J, Velasco A, Hage FG, Reyes E (2017) Comparison of ESC and ACC/AHA guidelines for myocardial revascularization. J Nucl Cardiol 24(3):1046–1053. https://doi.org/10.1007/s12350-017-0811-5

52. Endzeliņš E, Berger A, Melne V, Bajo-Santos C, Sobolevska K, Abols A et al (2017) Detection of circulating miRNAs: comparative analysis of extracellular vesicle-incorporated miRNAs and cell-free miRNAs in whole plasma of prostate cancer patients. BMC Cancer 17(1):730. https://doi.org/10.1186/s12885-017-3737-z

53. D'Alessandra Y, Devanna P, Limana F, Straino S, Di Carlo A, Brambilla PG et al (2010) Circulating microRNAs are new and sensitive biomarkers of myocardial infarction. Eur Heart J 31(22):2765–2773. https://doi.org/10.1093/eurheartj/ehq167

54. Kuwabara Y, Ono K, Horie T, Nishi H, Nagao K, Kinoshita M et al (2011) Increased microRNA-1 and microRNA-133a levels in serum of patients with cardiovascular disease indicate myocardial damage. Circ Cardiovasc Genet 4(4):446–454. https://doi.org/10.1161/CIRCGENETICS.110.958975

55. Rodriguez JA, Orbe J, Saenz-Pipaon G, Abizanda G, Gebara N, Radulescu F et al (2018) Selective increase of cardiomyocyte derived extracellular vesicles after experimental myocardial infarction and functional effects on the endothelium. Thromb Res 170:1–9. https://doi.org/10.1016/j.thromres.2018.07.030

56. Loyer X, Zlatanova I, Devue C, Yin M, Howangyin KY, Klaihmon P et al (2018) Intra-cardiac release of extracellular vesicles shapes inflammation following myocardial infarction. Circ Res 123(1):100–106. https://doi.org/10.1161/CIRCRESAHA.117.311326

Pharmacological Targets of Hypertension

Haaris Rahim, Yasmin Bashir, and Michael Schachter

© Springer Nature Switzerland AG 2019
C. Terracciano, S. Guymer (eds.), *Heart of the Matter*, Learning Materials in Biosciences,
https://doi.org/10.1007/978-3-030-24219-0_15

What You Will Learn in This Chapter

This chapter will begin by briefly explaining the parameters that form blood pressure, in addition to the cardiovascular burden of hypertension (HTN). From there, the key physiological modulators of blood pressure, namely autonomic tone, humoral factors, local and cellular factors, will be discussed, as will the renin–angiotensin–aldosterone system. Pharmacological agents used in the treatment of HTN will then be expanded upon, with the scientific rationale for each explored. Finally, we will explore two classes of drug that present new opportunities for the therapeutic modulation of blood pressure.

Learning Objectives
- Explain the key pathways and mediators of the renin–angiotensin–aldosterone (RAAS) system.
- Discuss the scientific rationale of current treatments for HTN, quoting at least five drug classes.
- Evaluate the potential use of neprilysin inhibitors and endothelin antagonists in HTN.

15.1 Hypertension

The term 'hypertension' refers to a persistent elevation of arterial blood pressure, with the 2017 American College of Cardiology and American Heart Association guidelines defining this as a blood pressure ≥ 130/80 mmHg [1]. This umbrella term can be subdivided into categories, with ~95% of hypertension (HTN) cases not arising secondary to another disease process, referred to as 'essential hypertension' [2]. With no singular cause, essential HTN has a multifaceted aetiology of complex polygenic factors, with superimposed environmental influences such as excess dietary sodium intake, obesity, alcohol consumption and low birth weight [2]. The remaining cases derive from an underlying primary cause and are therefore classified as 'secondary hypertension'. Known causes of secondary HTN include pre-eclampsia, renal artery stenosis, renal parenchymal disease, Conn's syndrome, phaeochromocytoma, use of the oral contraceptive pill and a panoply of rare genetic disorders [2].

Hypertension is a major risk factor for both cardiovascular and cerebrovascular disease, with critical interplay in the pathogenesis of atherosclerosis [3]. Longitudinal pathological outcomes of HTN include coronary heart disease (CHD), peripheral vascular disease (PVD), heart failure (HF), atrial fibrillation (AF), dissecting aneurysms and renal vascular damage [3]. Moreover, HTN constitutes the foremost modifiable risk factor for the development of strokes and transient ischaemic attacks, in addition to precipitating insidious damage over time via small vessel disease and vascular dementia [3].

A long-term reduction of diastolic BP by 5–6 mmHg is associated with a 35–40% reduction in the occurrence of stroke and a 20–25% reduction in the occurrence of CHD [4]. Furthermore, it has been noted that the reduction of BP via any regimen is more important for the prevention of cardiovascular events than the use of one drug class over another, and that larger reductions in BP are associated with larger reductions in risk [5]. Finally, BP reduction significantly reduces vascular risk across all individuals, irrespective of their baseline BP or comorbidities, providing strong support for the therapeutic reduction of BP in individuals with a history of cardiovascular (CV) disease, stroke, diabetes, or chronic kidney disease [6].

15

15.1.1   Regulation of Blood Pressure: The Fundamentals

Physiological control of BP is ultimately determined by two factors: cardiac output (CO) and peripheral vascular resistance (PVR). This is expressed in the following equation, where MBP is
'mean blood pressure' – the average blood pressure over a single cardiac cycle:

$$Mean\ Blood\ Pressure\,(MBP) = CO \times PVR$$

PVR represents the total resistance against which blood must be pumped, also known as 'afterload', and is primarily determined by the extent of vasoconstriction in the small arteries and arterioles. Cardiac output is the volume of blood ejected from the left ventricle over the course of a minute, governed by the heart rate (HR) and stroke volume (SV) in the following relationship:

$$Cardiac\ Output = HR \times SVR$$

Therefore, any variable that increases the HR, SV or degree of vasoconstriction will serve to increase blood pressure.
Physiologically, this is influenced by four factors:

1. *Autonomic Tone*: the balance between sympathetic and parasympathetic activation.
2. *Circulating Humoral Factors*: adrenaline, noradrenaline (NA), angiotensin II (Ang II), aldosterone and vasopressin (also referred to as 'anti-diuretic hormone/ADH').
3. *Local Regulators*: nitric oxide (NO), adenosine, endothelin (ET), eicosanoids, hydrogen ions, oxygen and myogenic factors.
4. *Cellular Factors*: intracellular calcium (Ca^{2+}).

Pharmacological agents represent an integral facet of BP reduction, however aren't the first line treatment. Rather, the initial intervention in any newly diagnosed essential HTN patient is CV risk factor reduction through the encouragement of lifestyle modifications [7]. These may include: reducing dietary sodium intake, weight loss to a BMI ≤ 25 kg/m^2, increasing physical activity, minimising alcohol consumption and quitting smoking [7].

15.2   Diuretics

Diuretics are agents that induce *diuresis*, the production of urination, and *natriuresis*, the excretion of sodium into urine. Through the increased loss of water from the body in the form of urine, blood volume and thus cardiac output decrease, reducing blood pressure. Importantly however, this is not the primary mechanism by which diuretics appear to exert their anti-hypertensive effect, with chronic diuretic use longitudinally associated with a restoration of plasma volume to slightly less than that of pre-treatment levels [8]. In fact, the precise system by diuretics facilitate a reduction in peripheral resistance is yet to be fully elucidated. Three main classes of diuretic exist for the management of HTN.

15.2.1 Thiazide Diuretics

A 'thiazide' is any diuretic agent that acts to inhibit the coupled reabsorption of sodium (Na^+) and chloride (Cl^-) ions at Na^+-Cl^- symporters in the distal convoluted tubule (DCT) of the nephron [9].

In the purest sense however, 'true thiazides' derive from the bicyclic heterocyclic benzene derivative, *benzothiadiazine*, including bendroflumethiazide and hydrochlorothiazide. That said, a second group of structurally distinct compounds known as 'thiazide-like' diuretics also generally fall in this group, including indapamide and chlorthalidone.

By increasing the removal of water and sodium from the body, thiazides play a vital role in the treatment of oedema, especially in heart and liver failure. Interestingly, a 2017 meta-analysis concluded that thiazide-like diuretics are in fact superior to classical thiazides in reducing BP, and typically result in fewer side effects associated with higher doses of thiazide diuretics [10]. This is of particular note given the positive correlation between the concentration of thiazide and the incidence of associated side effects, some of which include hyponatraemia, hypokalaemia, hyperglycaemia, hyperlipidaemia and impotence [11].

15.2.2 Loop Diuretics

Loop diuretics are predominantly used in the treatment of fluid overload, often in HF, and exert their effect in the thick ascending limb of the loop of Henle. Loop diuretics compete for Cl^- binding sites at Na^+-K^+-$2Cl^-$ (NKCC2) symporters, thus inhibiting the reabsorption of sodium and chloride ions whilst impairing a key urinary concentrating mechanism [12].

Their effect on reducing peripheral vascular resistance is more restricted, and as such they are not as widely used for the reduction of chronic HTN [13]. Furthermore, loop diuretics cause large changes in plasma volume and electrolyte status (e.g. hypokalaemia) without a significant effect on BP reduction [13]. Examples of loop diuretics include furosemide and bumetanide.

15.2.3 Potassium (K^+) – Sparing Diuretics

Though these will be discussed later in the chapter at greater length, potassium (K^+)-sparing diuretics principally act by inhibiting the active reabsorption of Na^+ in the late distal convoluted tubule and collecting duct, with net effects of modest diuresis and reduced secretion of K^+ and H^+ [14]. This 'potassium-sparing' effect increases the risk of hyperkalaemia, especially when co-administered with angiotensin-converting enzyme (ACE) inhibitors or angiotensin receptor blockers (ARBs), and also in patients with renal failure [14]. Examples include spironolactone and amiloride.

15.3 The Renin–Angiotensin–Aldosterone System (RAAS)

The renin–angiotensin–aldosterone system (RAAS) is a multifaceted neurohormonal signalling nexus chiefly responsible for the modulation of blood pressure and circulating fluid volume in humans [15]. The RAAS system is initiated by the release of renin from the *juxtaglomerular apparatus* in the kidney: a specialised cluster of modified pericytes within the glomerular capillaries that is highly sensitive to changes in Na^+ concentration and renal perfusion pressure [15]. Three main factors stimulate the release of renin into the blood [15]:

1. Low afferent arteriole pressure
2. Decreased NaCl delivery to the macula densa
3. Sympathetic stimulation of β_1-adrenergic receptors

Circulating renin catalyses the conversion of angiotensinogen – released from the liver – to angiotensin I (Ang I). The action of angiotensin-converting enzyme (ACE) in the pulmonary and renal endothelium subsequently catalyses the conversion of Ang I to angiotensin II (Ang II), with the resulting Ang II extensively binding to receptors on a host of body tissues [16]. This includes AT_1 receptors, which mediate a range of effects that summate to increase blood pressure [16], some of which summarised in ▢ Table 15.1. Of particular note is the binding of Ang II to AT_1 receptors on the adrenal cortex, stimulating the secretion of aldosterone, a mineralocorticoid hormone vital in the retention of sodium and water as part of the physiological response to a decrease in circulating fluid volume [16].

Aldosterone is released from the zona glomerulosa of the adrenal cortex and primarily exerts its effects via genomic mechanisms [17]. As a steroid hormone, it is able to pass through the cell membrane and translocate to the nucleus where it binds the hormone response element of specific genes, upregulating the expression of SGK1 (serine/threonine-protein kinase) and CHIF (channel-inducing factor) amongst others [17]. The net effect of this is twofold:

1. Upregulated expression of epithelial sodium (ENaC) channels in the apical membrane of the distal convoluted tubule and collecting duct, increasing the uptake of Na^+ from the renal tubule [17].
2. Upregulated expression of Na^+/K^+ ATPase enzymatic pumps in the basolateral membrane, increasing the absorption of sodium into the blood in exchange for excreted potassium [17].

The RAAS system can be targeted at multiple points using pharmacological agents, a selection of which we will now discuss.

▢ **Table 15.1** A summary of the key cardiovascular effects of Ang II-ET_1 receptor signalling at various sites around the body

Tissue	Summary of Ang II-ET_1 effect
Adrenal cortex	Triggers the release of aldosterone
Blood vessels	Vasoconstriction, endothelin release, pro-inflammatory
Kidney	Sodium and water retention, efferent glomerular arteriole constriction, glomerular and interstitial fibrosis
Heart	Coronary vasoconstriction, myocardial fibrosis, cellular hypertrophy, positive inotropy, pro-apoptotic, pro-inflammatory
Brain	Vasopressin secretion, sympathetic activation

15.3.1 Beta (β)-Blockers

The anti-hypertensive effects of beta (β)-blockers derive from their inhibition of β_1-adrenergic receptors, which are primarily expressed in the myocardium and in juxtaglomerular cells of the kidney (as well as the gastrointestinal tract) [18]. By inhibiting the sympathetic innervation of these areas, β-blockers reduce the heart rate whilst simultaneously decreasing the release of renin in the kidneys to downregulate RAAS activity [18]. Not all β-blockers are selective for the β_1-adrenergic receptor, with several classes available for the management of hypertension.

Notable side effects associated with β-blockers include bradycardia, dizziness, diarrhoea and hyperglycaemia [19], with the latter related to a reduction in insulin secretion observed with β-blocker usage [20]. Through this mechanism, the risk of diabetes is elevated [20]. Moreover, β-blockers are contraindicated in asthma due to the role of β_2-adrenergic receptors in the relaxation of smooth muscle in the airways [21]. With β_1-selective blockers exhibiting partial antagonism at β_2-receptors, the drugs increase the risk of bronchoconstriction.

15.3.2 Renin Inhibitors

The first effective oral renin inhibitor, Aliskiren, functions by binding to the renin S1/S3 pocket and another sub-pocket of the enzyme, inhibiting its activity [22]. However, with several studies identifying no significant efficacy advantage of Aliskiren over cheaper anti-hypertensive agents [23], its clinical utilisation has declined in recent years and therefore its role in future therapy remains uncertain.

15.3.3 Angiotensin-Converting Enzyme (ACE) Inhibitors

Angiotensin-converting enzyme (ACE) inhibitors are easily identifiable by the suffix '-pril', e.g. captopril, ramipril, enalapril, and function by inhibiting the ACE-mediated conversion of Ang I to Ang II, thus diminishing the effects of Ang II and reducing BP [24]. Logically, the inhibition of ACE would be assumed to result in a complete cessation in the production of Ang II. However, this in fact does not occur due to the so-called 'ACE escape' phenomenon, whereby the production of Ang II shifts to alternative pathways, blunting the reduction achieved with ACE inhibitors in a minority of patients [25]. For example, the aforementioned conversion reaction can also be catalysed by chymase enzymes, which is similarly renin-dependent.

Despite this, clinical trials have found ACE inhibitors to be highly effective for the reduction of BP, meaning they largely remain the first line drug agent for the management of essential HTN [24]. A common side effect of ACE inhibitors is a persistent dry cough, which occurs secondary to the accumulation of bradykinin [24]. ACE has an additional function as the catalyst for the breakdown of this inflammatory mediator which is thus inhibited. The persistent dry cough may also occur due to the increased activity of other vasoactive peptides yet to be fully described.

Other side effects include functional renal impairment, hypotension and hyperkalaemia, which is worsened when co-administered with potassium-sparing diuretics [26]. The side effects of ACE inhibitors are unrelated to the dose, in contrast to the efficacy which is proportional to concentration. ACE inhibitors are contraindicated in pregnancy as they increase the risk of foetal damage in the second and third trimester [26]. Other salient contraindications include individuals with a history of angioedema and bilateral renal artery stenosis.

15.3.4 Angiotensin Receptor Blockers (ARBs)

Angiotensin receptor blockers (ARBs) share the common suffix '-sartan', e.g. losartan, valsartan, and irbesartan. Similar to ACE inhibitors, they inhibit the functional effects of Ang II mediated by its binding to receptors around the body, albeit by an alternative mechanism with a more favourable side effect profile.

As the name suggests, ARBs antagonise the angiotensin II subtype 1 (AT1) receptor present on vascular smooth muscle, the adrenal cortex and the kidney, causing vasodilation and increased diuresis [27]. Synergistically, ARBs also hyper-stimulate AT2 receptors, a functional antagonist of the AT1 receptor, enhancing its vasodilatory effects. The current evidence base suggests that ACE inhibitors and ARBs have a similar blood pressure-dependent effect on decreasing the risk of coronary heart disease, stroke and HF [28, 29]. Interestingly however, a 2004 meta-analysis previously identified ARBs as being inferior to ACE inhibitors in the context of mortality reduction, with only the latter demonstrating a significant mortality relative risk reduction [30]. A larger meta-analysis of over 55,000 patients by Strauss and Hall also concluded that no ARBs had no discernable effect on overall mortality [31].

It was initially hypothesised that a double hit to the RAAS system by a combination therapy of ACE inhibitors and ARBs would be more beneficial than either drug alone at reducing the risks of hypertension [32]. However, upon examination of the meta-analysis reaching this conclusion, the findings were based upon a relatively small patient cohort ($n = 309$) from 10 studies. To greater investigate this claim, the multi-year ONTARGET trial randomised 25,620 patients to receive either an ACE inhibitor, ARB or both [33]. Two key conclusions were reached:

1. The ARB (telmisartan) was 'not inferior' to the ACE inhibitor (ramipril) in the primary outcomes of death from CV disease, myocardial infarction, stroke or hospitalisation for HF.
2. Combination therapy was associated with a marked increase in the risk of adverse effects including hypotension, renal dysfunction and hyperkalaemia, whilst also demonstrating no comparative benefit in the primary outcomes.

Thus, it is recommended to avoid combination therapy with ACE inhibitors and ARBs, except after coronary interventions [33]. This mirrored the result of the earlier VALIANT trial, which similarly found combination therapy as having no significant effect on the primary outcome while simultaneously increasing the incidence of hypotension [34].

15.3.5 Aldosterone Antagonists

Aldosterone antagonists (also known as *antimineralocorticoids*) block the binding of aldosterone to its receptors to induce natriuresis and diuresis via the inhibition of sodium reabsorption in the distal collecting tubule and collecting duct of the nephron [35]. This is far from the only effect of aldosterone however, as receptors are located throughout the body, predominately in the heart, vasculature, kidneys and brain. It is in the heart and vasculature where aldosterone receptor blockade has the greatest potential for reducing cardiovascular disease risk, with aldosterone signalling instigating myriad pathophysiological pathways including those for myocardial fibrosis and necrosis, vascular fibrosis, impaired fibrinolysis and endothelial dysfunction [35].

One of the foremost drugs of this class is spironolactone, a K^+-sparing diuretic commonly prescribed as an adjunct to other anti-hypertensive agents. This is evidenced by the RALES trial, which identified a 30% reduction in mortality over 3 years when spironolactone was added to standard antihypertensive treatment [36]. Furthermore, its implementation as an add-on therapy was reinforced by the ASCOT trial, which found that in HTN patients already taking an average of three BP-lowering agents, spironolactone facilitated a further reduction [37].

That said, due to the expansive distribution of mineralocorticoid and sex steroid receptors around the body, aldosterone antagonists are notably associated with several unpleasant side effects, notably gynecomastia and impotence in men, menstrual irregularities in women, and hyperkaliemia in both, thus predisposing to fatal ventricular arrhythmia [38].

A more selective aldosterone antagonist known as eplerenone was developed as a potential solution to this undesirable side effect profile. Eplerenone has a lower affinity for both the androgen and progesterone receptors when compared to standard aldosterone antagonists, however it also exhibits reduced binding with its pharmacological target, the mineralocorticoid receptor [38]. Overall however, with the risk of hyperkalaemia a significant and intrinsic limitation of this drug class, there exists much debate as to which mineralocorticoid receptor blocker, if any, is the optimal choice as an add-on therapy.

15.3.6 Calcium (Ca^{2+}) Channel Blockers (CCBs)

Calcium (Ca^{2+}) is an essential component of the cross-bridge cycle that underlies contraction within any muscular tissue. Increases in intracellular Ca^{2+} within vascular smooth muscle cells therefore results in vasoconstriction and increased peripheral vascular resistance [39]. By binding to multiple domains of the L-type calcium channel, CCBs block the transmembrane influx of Ca^{2+}, reducing vascular tone and thus lowering BP [39]. CCBs can be broadly classified into two groups:

- *Dihydropyridines*: For example nifedipine, amlodipine, lacidipine
- *Non-Dihydropyridines*: For example verapamil, diltiazem

The key difference between the two is that dihydropyridines are highly selective for the peripheral vascular smooth muscle cells while non-dihydropyridines also act on cardiomyocytes to exert a negative inotropic effect, i.e. reduce contractility [40]. This factor helps explain some of the side effects of the two classes. For example, dihydropyridines will stimulate a baroreceptor reflex due to the decrease in vascular tone, leading to increased heart rate and hence palpitations [39]. Conversely, non-dihydropyridines reduce cardiac output and can result in bradycardia and constipation, with vasodilatory side effects such as ankle oedema, flushing and headache tending to be milder compared to dihydropyridines [41]. Regardless, in both variants the efficacy and potential for side effects positively correlates with dose [41].

15.3.7   Alpha (α)-Blockers

Alpha 1 (α_1)-adrenergic receptors are a receptor subclass essential in the contraction of vascular smooth muscle, with the binding of adrenaline and noradrenaline activating an intracellular signalling cascade that facilitates a rapid increase in cytoplasmic Ca^{2+}. Thus, α_1-AR receptor blockade serves as a mechanism by which vasoconstriction may be inhibited. Three main classes exist:
1. Selective α_1 blockers: For example prazosin, doxazosin, terazosin
2. Non-selective α_1 and α_2 blockers: For example phenoxybenzamine, phentolamine
3. Combined α and β blockers: For example labetalol, carvedilol

α-blockers are no longer recommended as first line antihypertensive agents, however are still suitable as add-on agents to mainstay drugs such as ACE inhibitors, ARBs and CCBs [42]. This primarily derives from the ALLHAT trial, where doxazosin was associated with an elevated risk of CV events when compared to a thiazide-like diuretic (chlorthalidone) [42]. Side effects of α-blockers include dizziness, headache, palpitations and peripheral oedema [43]. Moreover, this particular antihypertensive is also associated with marked orthostatic hypotension, which increases the risk of falls and hip fractures in the elderly [43].

15.3.8   Centrally Acting Drugs

Centrally acting drugs (CADs) are designed to reduce sympathetic outflow by targeting the central nervous system. Thus far, CADs have taken the form of agonists of the α_2-adrenoceptor, imidazoline-1 receptor or both in the case of clonidine [44]. Agonism of these receptors facilitates a reduction in vasoconstriction and subsequently BP. Examples include [44]:
- *Methyldopa*: an α_2-agonist indicated for the management of HTN during pregnancy primarily due to its noted safety for both mother and fetus.
- *Clonidine*: a potent antihypertensive associated with a multitude of side effects, most notably rebound hypertension upon cessation of the drug, even if only a few doses are missed.

The selective imidazoline-1 receptor moxonidine on the other hand has very few side effects, however clinical trials found it failed to match other anti-hypertensives in terms of efficacy [44]. As such, it is now only used when other medications have already been tried and found unsuitable.

15.4   Where We're Heading: Novel Antihypertensives

15.4.1   Neprilysin Inhibition

The Natriuretic Peptide (NP) family is composed of five members, all of which act in opposition to the RAAS system. Five major categories of natriuretic peptide exist:
- Atrial Natriuretic Peptide (ANP)
- Brain Natriuretic Peptide (BNP)

- C-type Natriuretic Peptide (CNP)
- Dendroaspis Natriuretic Peptide (DNP)
- Urodilatin

ANP and BNP are released by the heart in response to atrial and ventricular stretch respectively, whilst CNP is released by the endothelium in response to shear stress and cytokines [45]. Both ANP and BNP bring about vasodilation and natriuresis, thus reducing BP [45]. The metabolism and clearance of natriuretic peptides occurs via two routes, the primary one being neprilysin (NEP)-catalysed degradation. This enzyme is largely present on the brush border membrane of vascular smooth muscle and the renal tubules and is involved in breakdown of the three major NPs to varying extents (CNP > ANP > BNP) [45]. Neprilysin also degrades other peptide hormones including urodilatin, bradykinin and adrenomedullin. The other route of NP metabolism is via the NP clearance receptor (NPR-C), which has a greater affinity for ANP than BNP, a factor contributing to the shorter half-life of ANP.

An understanding of these two mechanisms has allowed for the development of drugs that potentiate the half-life of NPs, allowing their BP-reducing effects to persist. To date, this has taken the form of neprilysin inhibitors, which have been used in combination with ACE inhibitors and ARBs to produce 'vasopeptidase inhibitors'. The initial vasopeptidase inhibitor, omapatrilat, was a combined neprilysin inhibitor and ACE inhibitor. Whilst its efficacy was validated in both rat and human studies, both the OVERTURE and OCTAVE trials concluded it didn't outperform existing anti-hypertensives, yet had a significantly worse range of adverse effects [46, 47]. Of particular concern was the elevated incidence of angioedema, particularly in African-American individuals, which was ascribed to the double blockade of bradykinin breakdown normally metabolised by either neprilysin or ACE [47].

A novel approach to these issues came with the development of angiotensin receptor-neprilysin inhibitors (ARNIs), which comprise a neprilysin inhibitor and an ACE inhibitor in a 1:1 molecular mixture. Perhaps the most well-known example of this class is Entresto, a combination of sacubitril and valsartan [48]. In this way, the combination drug does not inhibit ACE directly.

The PARADIGM-HF compared Entresto against a commonly prescribed ACE inhibitor (enalapril) in HF patients with a reduced ejection fraction (HFrEF) [48]. Ultimately, the study was concluded early (concordant with 'prespecified rules') due to overwhelming evidence that Entresto had greater efficacy in reducing the risk of death and hospitalisation for HFrEF. This study contributed to the evidence base for the use of ARNIs in the treatment of heart failure, however given their underlying pharmacodynamics, this suggests potential value as antihypertensives. Clearly, further research is needed into the efficacy of ARNIs in the treatment of HTN.

15.4.2   Endothelin Antagonists

Endothelins are a group of three peptides – ET-1, ET-2 and ET-3 – of which ET-1 is the most abundant isoform in the cardiovascular system and a potent vasoconstrictor [49]. Endothelins act on the G_q-protein coupled receptors ETA and ETB found throughout the body, exhibiting equal affinity for both receptors. The exception to this is ET-3, which has a lower affinity for ETA receptors. ET-1 is continuously synthesised in vascular endothelial cells, with production further enhanced by stimuli including angiotensin II, aldosterone, growth factors and cytokines, in addition to certain pathological states such as hypertension and obesity [49]. ET-1 therefore potentiates RAAS system effects, contributing to vascular tone and cell proliferation [49].

The activation of ETA and ETB receptors on endothelial and smooth muscle cells by ET-1 instigates a signalling cascade that increases intracellular IP_3 and DAG second messengers. Within smooth muscle cells, this facilitates an increase in intracellular Ca^{2+} with subsequent vasoconstriction and cell proliferation. Conversely, the binding of ET-1 to endothelial ETA receptors has an indirect vasodilatory effect via the upregulated production of vasodilators NO and prostacyclin [49]. That said, the overall effect of ET-1 however, is to increase BP.

Interestingly, ET-1 has also been recognised as a key mediator in the development of pulmonary artery hypertension (PAH), a rare disease characterised by the narrowing of pulmonary arterioles [49]. In PAH, elevated levels of ET-1 are present in the plasma, coupled with an increase in its vasoconstrictive capacity.

Mechanistically, the latter phenomenon arises because high ET-1 upregulates the expression of ETA and ETB receptors on smooth muscle cells, with a concomitant decrease in endothelial ETB expression, enhancing its vasoconstrictive activity [49]. Despite the profound effects of ET-1 on major organs and its key role in the development of cardiovascular disease, the only established indication for the use of endothelin receptor antagonists (ERAs) is in PAH and not for peripheral hypertension [49]. This is in part due to their high cost, but also to the elevated risk of liver toxicity associated with usage [49].

Take-Home Message

- Hypertension (HTN) is a major risk factor for both cardiovascular and cerebrovascular disease. It can be categorised into essential (no singular cause) and secondary (deriving from an underlying primary cause) subsets.
- Diuretics mediate the retention of sodium (and thus water) in the renal tubule to decrease blood volume and hence pressure.
- ACE inhibitors remain largely the first line pharmacological agent for essential HTN and inhibit the ACE-mediated conversion of angiotensin I to angiotensin II.
- The PARADIGM-HF study demonstrated the efficacy of angiotensin receptor-neprilysin inhibitors in the treatment of HFrEF, however further research is needed into their antihypertensive benefit. That said, they remain a promising novel therapy.

References

1. Whelton PK, Carey RM, Aronow WS, Casey DE, Collins KJ, Dennison Himmelfarb C, et al (2018) ACC/AHA/AAPA/ABC/ACPM/AGS/APhA/ASH/ASPC/NMA/PCNA guideline for the prevention, detection, evaluation, and management of high blood pressure in adults: executive summary. Hypertension. 71(6): 1269–1324
2. Carretero OA, Oparil S (2000) Essential hypertension. Part I: definition and etiology. Circulation 101:329
3. Gaciong Z, Siński M, Lewandowski J (2013) Blood pressure control and primary prevention of stroke: summary of the recent clinical trial data and meta-analyses. Curr Hypertens Rep 15:559

4. Collins R, Peto R, MacMahon S, Godwin J, Qizilbash N, Collins R et al (1990) Blood pressure, stroke, and coronary heart disease. Part 2, short-term reductions in blood pressure: overview of randomised drug trials in their epidemiological context. Lancet 335:827

5. Turnbull F, Neal B, Algert C, Chalmers J, Woodward M, MacMahon S et al (2003) Effects of different blood-pressure-lowering regimens on major cardiovascular events: results of prospectively-designed overviews of randomised trials. Lancet 362:1527

6. Ettehad D, Emdin CA, Kiran A, Anderson SG, Callender T, Emberson J et al (2016) Blood pressure lowering for prevention of cardiovascular disease and death: a systematic review and meta-analysis. Lancet 387:957

7. Chobanian AV, Bakris GL, Black HR, Cushman WC, Green LA, Izzo JL et al (2003) Seventh report of the joint national committee on prevention, detection, evaluation, and treatment of high blood pressure. Hypertension 42:1206

8. Shah S, Khatri I, Freis ED (1978) Mechanism of antihypertensive effect of thiazide diuretics. Am Heart J 95:611

9. Hropot M, Fowler N, Karlmark B, Giebisch G (1985) Tubular action of diuretics: distal effects on electrolyte transport and acidification. Kidney Int 28:477

10. Liang W, Ma H, Cao L, Yan W, Yang J (2017) Comparison of thiazide-like diuretics versus thiazide-type diuretics: a meta-analysis. J Cell Mol Med 21:2634

11. Salvetti A (2006) Thiazide diuretics in the treatment of hypertension: an update. J Am Soc Nephrol 17:S25

12. Wittner M, Di Stefano A, Wangemann P, Greger R (1991) How do loop diuretics act? Drugs 41:1

13. Tamargo J, Segura J, Ruilope LM (2014) Diuretics in the treatment of hypertension. Part 2: loop diuretics and potassium-sparing agents. Expert Opin Pharmacother 15:605

14. Wile D (2012) Diuretics: a review. Ann Clin Biochem 49:419

15. Peti-Peterdi J, Harris RC (2010) Macula Densa sensing and signaling mechanisms of renin release. J Am Soc Nephrol 21:1093

16. Kawai T, Forrester SJ, O'Brien S, Baggett A, Rizzo V, Eguchi S (2017) AT1 receptor signaling pathways in the cardiovascular system. Pharmacol Res 125:4

17. Soundararajan R, Pearce D, Ziera T (2012) The role of the ENaC-regulatory complex in aldosterone-mediated sodium transport. Mol Cell Endocrinol 350:242

18. Weber MA (2005) The role of the new β-blockers in treating cardiovascular disease. Am J Hypertens 18:169

19. Barron AJ, Zaman N, Cole GD, Wensel R, Okonko DO, Francis DP (2013) Systematic review of genuine versus spurious side-effects of beta-blockers in heart failure using placebo control: recommendations for patient information. Int J Cardiol 168:3572

20. Gress TW, Nieto FJ, Shahar E, Wofford MR, Brancati FL (2000) Hypertension and antihypertensive therapy as risk factors for type 2 diabetes mellitus. N Engl J Med 342:905

21. Morales DR, Guthrie B, Lipworth BJ, Donnan PT, Jackson C (2011) Prescribing of β-adrenoceptor antagonists in asthma: an observational study. Thorax 66:502

22. Müller DN, Luft FC (2006) Direct renin inhibition with aliskiren in hypertension and target organ damage. Clin J Am Soc Nephrol: CJASN 1:221

23. Gao D, Ning N, Niu X, Wei J, Sun P, Hao G (2011) Aliskiren vs. angiotensin receptor blockers in hypertension: meta-analysis of randomized controlled trials. Am J Hypertens 24:613

24. Messerli FH, Bangalore S, Bavishi C, Rimoldi SF (2018) Angiotensin-converting enzyme inhibitors in hypertension: to use or not to use? J Am Coll Cardiol 71:1474

25. MacFadyen RJ, Lee AFC, Morton JJ, Pringle SD, Struthers AD (1999) How often are angiotensin II and aldosterone concentrations raised during chronic ACE inhibitor treatment in cardiac failure? Heart 82:57

26. Sica DA (2004) Angiotensin-converting enzyme inhibitors side effects--physiologic and non-physiologic considerations. J Clin Hypertens (Greenwich) 6:410

27. Taylor AA, Siragy H, Nesbitt S (2011) Angiotensin receptor blockers: pharmacology, efficacy, and safety. J Clin Hypertens 13:677

28. Turnbull F, Neal B, Pfeffer M, Kostis J, Algert C, Woodward M et al (2007) Blood pressure-dependent and independent effects of agents that inhibit the renin-angiotensin system. J Hypertens 25:951

29. Munger MA (2011) Use of angiotensin receptor blockers in cardiovascular protection: current evidence and future directions. P T 36:22

15

30. Strippoli GFM, Craig M, Deeks JJ, Schena FP, Craig JC (2004) Effects of angiotensin converting enzyme inhibitors and angiotensin II receptor antagonists on mortality and renal outcomes in diabetic nephropathy: systematic review. Br Med J 329:828

31. Strauss MH, Hall AS (2006) Angiotensin receptor blockers may increase risk of myocardial infarction unraveling the ARB-MI paradox. Circulation 114:838

32. Jennings DL, Kalus JS, Coleman CI, Manierski C, Yee J (2007) Combination therapy with an ACE inhibitor and an angiotensin receptor blocker for diabetic nephropathy: a meta-analysis. Diabet Med 24:486

33. Investigators TO (2008 [cited 2019 Jan 3]) Telmisartan, ramipril, or both in patients at high risk for vascular events. N Engl J Med [Internet] 358(15):1547–59. Available from: http://www.nejm.org/doi/abs/10.1056/NEJMoa0801317

34. Pfeffer MA, McMurray JJV, Velazquez EJ, Rouleau J-L, Køber L, Maggioni AP et al (2003) Valsartan, captopril, or both in myocardial infarction complicated by heart failure, left ventricular dysfunction, or both. N Engl J Med 349:1893

35. Guichard JL, Clark D, Calhoun DA, Ahmed MI (2013) Aldosterone receptor antagonists: current perspectives and therapies. Vasc Health Risk Manag 9:321–31

36. Pitt B, Zannad F, Remme WJ, Cody R, Castaigne A, Perez A et al (1999) The effect of spironolactone on morbidity and mortality in patients with severe heart failure. Randomized Aldactone Evaluation Study Investigators. N Engl J Med 341:709

37. Chapman N, Dobson J, Wilson S, Dahlöf B, Sever PS, Wedel H et al (2007) Effect of spironolactone on blood pressure in subjects with resistant hypertension. Hypertension 49:839

38. Brown NJ (2003) Eplerenone: cardiovascular protection. Circulation 107:2512

39. Muntwyler J, Follath F (2001) Calcium channel blockers in treatment of hypertension. Prog Cardiovasc Dis 44(3):207–216

40. Opie LH (1997) Pharmacological differences between calcium antagonists. Eur Heart J 18:71

41. Hedner T (1986) Calcium channel blockers: spectrum of side effects and drug interactions. Acta Pharmacol Toxicol (Copenh) 58(Suppl 2):119–130

42. ALLHAT Officers and Coordinators for the ALLHAT Collaborative Research Group. The Antihypertensive and Lipid-Lowering Treatment to Prevent Heart Attack Trial (2002) Major outcomes in high-risk hypertensive patients randomized to angiotensin-converting enzyme inhibitor or calcium channel blocker vs diuretic: The Antihypertensive and Lipid-Lowering Treatment to Prevent Heart Attack Trial (ALLHAT). JAMA 288:2981. https://jamanetwork.com/journals/jama/fullarticle/195626

43. Carruthers SG (1994) Adverse effects of α1-adrenergic blocking drugs. Drug Saf 11:12

44. Sica DA (2007) Centrally acting antihypertensive agents: an update. J Clin Hypertens (Greenwich) 9:399

45. Bavishi C, Messerli FH, Kadosh B, Ruilope LM, Kario K (2015) Role of neprilysin inhibitor combinations in hypertension: insights from hypertension and heart failure trials. Eur Heart J 36:1967

46. Packer M, Califf RM, Konstam MA, Krum H, McMurray JJ, Rouleau JL et al (2002) Comparison of omapatrilat and enalapril in patients with chronic heart failure: the omapatrilat versus enalapril randomized trial of utility in reducing events (OVERTURE). Circulation 106:920

47. Kostis JB, Packer M, Black HR, Schmieder R, Henry D, Levy E (2004) Omapatrilat and enalapril in patients with hypertension: the omapatrilat cardiovascular treatment vs. Enalapril (OCTAVE) trial. Am J Hypertens 17:103

48. McMurray JJV, Packer M, Desai AS, Gong J, Lefkowitz MP, Rizkala AR et al (2014) Angiotensin–neprilysin inhibition versus enalapril in heart failure. N Engl J Med 371:993

49. Schiffrin EL (2005) Vascular endothelin in hypertension. Vasc Pharmacol 43:19

The Coagulation Cascade and its Therapeutic Modulation

Lieze Thielemans, Moghees Hanif, and James Crawley

© Springer Nature Switzerland AG 2019
C. Terracciano, S. Guymer (eds.), *Heart of the Matter*, Learning Materials in Biosciences,
https://doi.org/10.1007/978-3-030-24219-0_16

What You Will Learn in This Chapter

This chapter will provide a comprehensive overview of both primary and secondary haemostasis at a gross and molecular level, in addition to examining the relevant signalling and cellular mediators of these dual processes. We will then discuss three key regulatory mechanisms, whilst simultaneously exploring the clinical relevance of both the aforementioned mechanisms and the aberrance of coagulation.

Learning Objectives
- Be able to provide a comprehensive overview of both primary and secondary haemostasis, including the intrinsic and extrinsic initiation pathways.
- Appreciate the structural importance of serine processes and the interplay this has with their functional activity.
- Explain the importance and action mechanisms of TFPI, Activated Protein C and Antithrombin.

16.1 Haemostasis: The Fundamentals

The term 'haemostasis' derives as a portmanteau of 'haemo' and 'stasis', literally referring to the 'stopping of blood', and is a vital apparatus of maintaining haemodynamic stability through the prevention of blood leaving a damaged blood vessel. Haemostasis exists as a balance between localised coagulation at the site of vascular injury and maintenance of normal blood flow elsewhere in the circulation, achieved through a carefully controlled balance of procoagulant and anticoagulant mediators. By ensuring that responses are both localised and transient, haemostasis is able to simultaneously prevent thrombotic events whilst also allowing the restoration of patency at the site of vascular injury. This primary platelet plug is established by primary haemostasis, which provides a surface for the coagulation cascade to stabilise the platelet plug, ultimately creating a stable clot through secondary haemostasis [1].

16.1.1 Vascular Injury and Platelets

Under physiological conditions, the vascular endothelium establishes an anticoagulant luminal surface through the expression of a variety of compounds including nitric oxide (NO) and heparin-like glycosaminoglycans (GAGs). In the event of vascular disruption (e.g. from injury), several procoagulant compounds in the underlying subendothelial matrix are exposed to the blood flow, creating a naturally procoagulant surface and activating a dormant yet always-present thrombotic apparatus. Initially, intravascular signalling molecules released from activated endothelial cells (e.g. endothelin), coupled with activated platelet-derived thromboxane A2 (TXA2), diffuse to the underlying smooth muscle layer, triggering contraction of smooth muscle cells and thereby locally constricting the vessel to prevent haemorrhage [2].

Platelets, also known as thrombocytes, are anucleate biconvex fragments of cytoplasm derived from megakaryocytes that circulate inactive in the plasma due to nitric oxide (NO) and prostacyclin (PGI2) released from the endothelium. Following vascular disruption, platelets interact with the newly exposed subendothelial collagen and von Willebrand factor (vWF) [3].

16.1.2   von Willebrand Factor (vWF)

von Willebrand factor (vWF) is a multimeric glycoprotein synthesised in both endothelial cells and megakaryotes, and has two major functions:
1. Forms an adhesive bridge between the damaged vessel wall and platelets, vital for primary haemostasis.
2. Stabilises and protects factor VIII (FVIII).

During translation, vWF is arranged into dimers that subsequently form multimers through the establishment of disulfide bridges with other dimers. Bond formation is facilitated by the isomerase activity of the propeptide sequence that is later cleaved to produce the highly reactive mature vWF multimer. In endothelial cells, vWF undergoes one of three fates:
1. Stored in intracellular Weibel-Palade bodies to be released if activated.
2. Stored in Weibel-Palade bodies and constitutively released in low concentrations.
3. Secreted directly by the endothelial cells into the subendothelial matrix.

If stored in the Weibel-Palade storage granules, vWF is compacted into a globular form that changes upon release due to environmental stimuli, namely pH. Upon release into the circulation, the highly reactive multimers are cleaved at specific sites by the metalloprotease ADAMTS13 into smaller forms (up to 20MD a), reducing the number of available platelet binding sites and thus reactivity. Failure of this cleavage reaction results in premature interactions with platelets due to increased binding affinity.

In contrast, platelet vWF originates from megakaryotes and is stored in α-granules to be released upon platelet activation. This pool does not contribute to plasma vWF levels, and therefore is not susceptible to ADAMTS13, proteolysis, until it is released [3].

16.2   Primary Haemostasis

The initial response to vessel injury is characterised by *primary haemostasis*. Upon vessel wall damage, the newly exposed subendothelial collagen catches on circulating globular vWF in plasma. Shear stress causes the immobilised vWF to unravel into a string-like configuration, now exposing the binding sites for platelet GPIbα in the A1 domain of vWF [4]. This facilitates the binding of platelets to vWF, enabling it to bind to collagen via GPIa-IIa and the GPVI complex. This binding process initiates a series of activation steps (via Phospholipase C-γ) within the platelet that ends in degranulation and the release of further vWF from α-granules, enabling further platelet aggregation via feedback activation and the formation of a primary haemostatic plug [5].

Platelet-released vWF has notably higher binding affinity for collagen and platelets than plasma vWF forms. Platelet activation is mediated by agonist action by bound platelet secreted products and local prothrombotic factors, where the GP VI interaction with collagen is thought to be the major signalling receptor involved in platelet activation on exposed collagen. The conformational change achieved by the association of GPVI and collagen results in platelet degranulation and release of α-granule contents, dense granule contents and release of TXA2 [6, 7].

16

- *α-Granule Contents*: integral membrane proteins (e.g. P-selectin, GPIIb/IIIa, GPIbα), procoagulant/anticoagulant/fibrinolytic proteins (e.g. fibrinogen, factor V, factor IX, TFPI, protein S, plasminogen), adhesion proteins (vWF and fibrinogen) and chemokines.
- *Dense Granule Contents*: ATP, ADP, serotonin, histamine, Ca^{2+} and polyphosphates.

Soluble agonists act on GPCRs. Released from both dense granules and damaged endothelial cells, ADP is one such example, binding to $P2Y_1$ and $P2Y_{12}$ to enhance platelet activation. Similarly, TXA2 released by stimulated platelets also serves as a second messenger by activating the TXA2 receptor. Critically, platelet activation triggers conformational changes to glycoprotein IIb-IIIa: an integrin complex that serves as a vital platelet receptor for both fibrinogen and von Willebrand factor. The binding of ADP instigates the formation of this receptor, promoting both vWF binding and platelet–platelet aggregation via fibrinogen linkage [4, 5].

16.2.1 Summary of Primary Haemostasis

1. Vessel wall damaged → subendothelial collagen exposed → vWF binds
2. Shear stress elongates vWF, exposing multiple binding sites
3. Shear stress reveals vWF A1 domain → catches on platelet GPIb
4. Platelet rolls along vWF via GPIb
5. GpIIb/IIIa adopts active configuration → 'fixes' platelet to vWF
6. Platelets degranulate, releasing ADP, thromboxane A2 more vWF
7. Platelet feedback completes activation
8. Fibrinogen links platelets via GpIIb/IIIa
9. Further platelets captured, forming platelet plug

16.2.2 Further Platelet Activation in Secondary Haemostasis

Although secondary haemostasis will be discussed later in the chapter, it is worth noting that further platelet activation by thrombin, the terminal serine protease of the coagulation cascade, occurs as it cleaves two GPCRs (PAR1 and PAR4). The resulting conformational changes activate these receptors, mediating further granule release and integrin activation. PAR1 is active at low thrombin concentrations, whereas PAR4 activates platelets at higher thrombin concentrations [5].

16.2.3 Clinical Relevance: Antiplatelet Therapies

Given the central role of platelets in primary haemostasis, it follows that excessive platelet activation and aggregation can be life threatening. Pharmacological manipulation of primary haemostasis is used to treat patients at high risk of thrombosis. Examples of prophylactic antiplatelet use include those patients with stable angina, atrial fibrillation or a high risk of stroke. The main classes of antiplatelets and their mechanisms of action are described below.

Aspirin is one of the most commonly used antiplatelet agents. Aspirin irreversibly inhibits COX1 enzymes within platelets. Inhibition of this enzyme decreases the platelet production of prostaglandins and TXA2. Platelet TXA2 production is key in subsequent platelet activation and aggregation, as previously detailed. Reducing TXA2 production serves to diminish platelet action in primary haemostasis. The inhibition of primary haemostasis in this way, decreases the blood's capacity to form clots, and thus lessens the thrombotic risk. This is a key concept in stroke prevention [8, 9].

A second class of antiplatelet drugs target the ADP receptor on the platelet surface in order to inhibit platelet aggregation. Clopidogrel and Prasugrel both work through irreversible competitive inhibition of P2Y$_{12}$ receptor. Consequently, fibrin receptor activation is decreased, inhibiting platelet aggregation. Clopidogrel is commonly used in the immediate treatment of myocardial infarction, however is also employed prophylactically to prevent MI or stroke. It is important to note that this is a prodrug and thus needs to be metabolised by the liver before it is biologically active. Factors affecting the liver metabolising enzyme CYP450, such as polymorphisms or other drugs, may therefore alter the active dose released in the circulation [10].

Other, more modern, antiplatelet agents are slowly being introduced into common clinical care. One such class are PAR1 antagonists, examples of which include Vorapaxar and Atopaxar. These drugs work via the competitive inhibition of the PAR1 on the platelet surface, thus decreasing thrombin-mediated platelet aggregation. Though they do seem to be associated with some reduction in risk of cardiovascular mortality, it is important to note that several clinical studies have associated PAR1 antagonists with an increased bleeding risk, limiting their usage in the clinical setting [11].

Finally, GPIIb/IIIa blockers (e.g. Abciximab and Eptifibatide) are monoclonal antibodies or small molecules, respectively, that antagonise the GPIIb/IIIa receptor on the platelet surface, thereby preventing the cross-linking of activated platelets by fibrinogen and thus platelet aggregation. Due to a limited evidence base, Abciximab is currently only used in patients undergoing percutaneous coronary intervention, however further trials are ongoing [12].

Antiplatelet drugs are greatly beneficial in the secondary prevention of coronary heart disease, stroke or acute coronary events. However, the effects in primary prevention are much less clear and clinicians must always bear in mind the balance between thrombotic risk vs. bleeding risk.

16.3 Secondary Haemostasis

With the primary haemostatic plug formed, the newly activated platelets exhibit increased intracellular calcium (via dense granules) and provide a highly negatively charged phospholipid surface environment that promotes the assembly of activated coagulation factors. Secondary haemostasis is a tightly regulated process consisting of multiple proteolytic reactions that leads to the formation of an insoluble cross-linked fibrin meshwork that reinforces and stabilises the primary platelet plug. This process is commonly referred to as the 'coagulation cascade', and comprises many different coagulation factors with diverse and specific roles [13].

16.3.1 Coagulation Factors

Secondary haemostasis is largely mediated by coagulation factors: serine protease enzymes that sequentially catalyse one another in a synergistic manner in order to exponentially

Zymogens (inactive)	Serine proteases	Cofactors	Inhibitors	
prothrombin*	thrombin	TF	TFPI	(Kunitz-type)
FVII*	FVIIa*	FVa	Protein C*	(serine protease)
FIX*	FIXa*	FVIIIa	Protein S*	(cofactor for APC)
FX*	FXa*		Antithrombin	(serpin)
FXI	FXIa			
FXII	FXIIa			
FXIII				

* = contains Gla domain

Fig. 16.1 A list of all the components necessary for the coagulation process to take place. Zymogens are inactive in circulation and are activated to serine proteases by the specific removal of an activation peptide when required. Only some factors contain the Gla domain. Co-factors TF, FVa and FVIIIa are crucial components to facilitate factor activation. The inhibitors are important to control and regulate the coagulation process

enhance the formation and cross-linking of fibrin at the primary haemostatic plug. Apart from tissue factor (TF), coagulation factors generally follow a roman numeral naming process, for example: factor VIII (FVIII), and are referred to in some textbooks as 'clotting factors'. The only factors that are not serine proteases are factors, FV, FVIII and FXIII. Whilst most pro- and anticoagulant factors are synthesised in the liver, coagulation factors circulate in the plasma as *zymogens*: inactive enzyme precursors, which are activated by the specific proteolytic removal of their activation peptide. This paradigm makes the factors able to be immediately recruited upon injury. Co-factors assist this enzymatic activation process and are vital for the amplification of coagulation factor production when needed. Figure 16.1 provides an overview of the various coagulation factors [14].

16.3.2 Coagulation Factor Constituents

Coagulation factors comprise distinct domains crucial to their function in the cascade. This includes the Gla domain, epidermal growth factor (EGF) domain, Kringle domain and the serine protease domain, which are shared among several factors. For example, Factors X, VII, IX and protein C share a similar domain organisation of one Gla domain, two EGF domains and one serine protease domain, thereby forming a homologous modular structure. Despite this structural similarity, all factors perform different functions (Fig. 16.2).

— *Serine Protease (SP) Domain*: provides the serine protease with its catalytic activity where it cleaves substrates after specific Arg and Lys residues. SP domains of all the haemostatic proteases come from the same family of proteases and contain a catalytic triad (His/Asp/Ser) that both defines the protease and imparts mutual specificity. Serine proteases are activated by the proteolytic removal of their activation peptide.

— *Epidermal Growth Factor (EGF) Domain*: involved in imparting binding affinity for both substrates and co-factors.

— *Kringle Domain*: this domain plays a role in binding mediators such as phospholipid membranes and other proteins. It is also involved in the regulation of proteolytic activities.

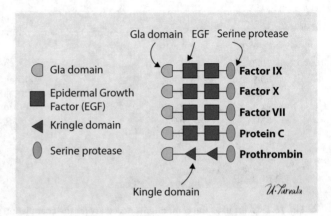

Fig. 16.2 Coagulation factors comprise four discrete domains: Gla, EGF, SP and the Kringle domain. The organisation of these determines their specific function, with most factors (including factors IX, X, VII, protein C and prothrombin) sharing a common organisation. Co-factors such as factors V and VIII have a different domain organisation

— *Gla Domain*: this small domain (35–40 amino acids long) imparts the capacity for clotting factors to bind to negatively charged phospholipid surfaces. The Gla domain ultimately comes to define vitamin K-dependent proteins.

When Gla-containing factors are synthesised, they possess a number of glutamic acid residues within their N-termini. These residues only have a single carboxylic acid moiety. However, after undergoing post-translational modification facilitated by vitamin K-dependent carboxylase, the residues are modified to γ-carboxyglutamic acid, with two carboxylic acid moieties. This provides Gla residues with a strong negative charge that imparts the ability to bind to calcium ions. By coordinating those ions within its structure, the domain is able to fold up into its functional conformation. Another structural component of the Gla domain is the Omega loop: a hydrophobic structure that enables factors to bind to hydrophobic phospholipid surfaces present on platelets [15].

16.4 Coagulation Cascade

The coagulation cascade has been traditionally classified into intrinsic and extrinsic pathways, both of which converge into a common pathway of Factor X activation. Most reactions occurring in the cascade require calcium ions due to the involvement of Gla-containing factors [16].

16.4.1 Extrinsic Pathway

Tissue Factor (TF) is a subendothelial integral membrane protein primarily responsible for the initiation of the extrinsic coagulation pathway through the formation of thrombin. Due to TF expression at extravascular sites not usually exposed to blood flow, it is able to act as an efficient trigger of the extrinsic pathway, exemplified by its enhanced presence in vital organs such as the brain and heart, conferring further haemostatic protection.

Tissue factor acts as a cellular receptor and cofactor for FVII and FVIIa, a small amount of which routinely circulates in the vasculature. Following vessel wall damage, TF is exposed to the blood. This potent activator binds to FVII/FVIIa, which is brought to the surface via Gla domains. This triggers an activating conformational change in the serine protease domain that facilitates the formation of a TF-FVIIa complex able to catalyse the proteolytic activation of FIXa and FXa. The newly formed Factor Xa is now able to facilitate the formation of thrombin from prothrombin via the removal of the Kringle and Gla domains in an inefficient process. Despite this low yield, thrombin is now able to indirectly augment its own production via the activation of two key co-factors: FVa and FVIIIa.

Factor VIIIa is now capable of binding with FIXa to form FVIIIa-FIXa complexes that enhance the production of FXa more so than TF-FVIIa, thereby providing an apparatus for the generation of FXa that is no longer reliant on TF. FVa acts as a co-factor to FXa, amplifying its activation of thrombin from prothrombin 200,000-fold and rapidly achieving an exponential increase in thrombin production. Finally, thrombin mediates the conversion of fibrinogen, a soluble structural protein, to insoluble fibrin, through the cleavage of fibrinopeptide A (FpA) and fibrinopeptide B (FpB). The resulting fibrin self-associates into a complex 3D matrix-like structure that stabilises the platelet plug.

Importantly, these reactions require Ca^{2+} and a phospholipid membrane (e.g. platelets), evidenced by the clinical utilisation of citrate to prevent the coagulation of blood. Citrate is able to collate Ca^{2+}, highlighting its importance (◘ Fig. 16.3).

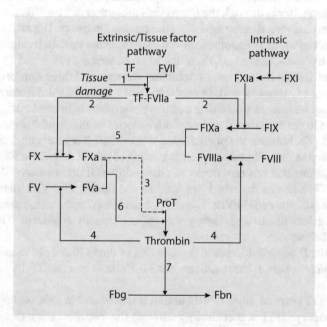

◘ **Fig. 16.3** A schematic representation of the coagulation cascade. (1) After injury, tissue factor (TF) binds to FVII/FVIIa, forming a TF-FVIIa complex. (2) The TF-FVIIa complex converts FIX to FIXa and FX to FXa. (3) FXa converts prothrombin to thrombin, generating small quantities of thrombin. (4) Thrombin activates two co-factors: FVIII and FV, forming FVIIIa and FVa, respectively. (5) FVIIIa acts as a co-factor for FIX and forms the FVIIIa-FIX complex that augments the production of FXa. (6) FVa acts as a co-factor for FXa to amplifying its function. The FVa-FXa complex converts pro-thrombin to thrombin, generating large quantities of thrombin. (7) Thrombin converts soluble fibrinogen to insoluble fibrin

16.4.2 Intrinsic Pathway

The intrinsic pathway is also known as the 'contact pathway', and exists as a parallel mechanism for the activation of thrombin. It can be initiated by various non-physiological agents such as platelet-secreted polyphosphate and the presence of extracellular DNA, with the latter a potential indicator of NETosis (neutrophil extracellular traps). The intrinsic pathway begins with the activation of Factor XI to FXIa. From this cleavage, FXIa is able to further activate Factor IX, with FIXa contributing to the activation of FX when forming a complex with its co-factor (Factor VIII).

16.5 Regulation of Coagulation

A number of physiological control mechanisms exist to ensure haemostasis remains localised to the area of vessel damage. The main regulators involved are tissue factor pathway inhibitor (TFPI), activated protein C and antithrombin.

16.5.1 Tissue Factor Pathway Inhibitor

Tissue factor pathway inhibitor (TFPI) is a single-chain polypeptide that arrests the initiation stage of coagulation through the irreversible inhibition of TF-FVIIa complexes in an FXa-dependent manner, thereby inhibiting the extrinsic pathway. The majority of TFPI is synthesised in the vascular endothelium and may therefore remain localised here or circulate freely in plasma. In humans, there exists two isoforms: TFPI-α and TFPI-β. TFPI-α is a 43 Da Kunitz-type serine protease inhibitor, consisting of three Kunitz domains (K1, K2 and K3) and a C-terminal part. In contrast, TFPI-β lacks the third Kunitz domain and C-terminal region, instead possessing a glucosylphosphatidylinositol anchor that facilitates adhesion to the endothelial surface. Both isoforms exhibit inhibitory action towards TF-FVIIa and FXa, however the physiological role of TFPI-β is poorly understood in vivo.

The first step in the TFPI pathway is the binding of the K2 domain to FXa, forming an FXa-TFPI complex that can now dock onto the endothelial surface near TF-FVIIa. Once docked, the TFPI Kunitz domain 1 can bind to and inhibit TF-FVIIa, manifesting as a large, inactive quaternary TF-FVIIa/FXa-TFPI complex. With the serine proteases of both coagulation factors inactivated, their procoagulant capacity is ablated. This process is summarised below:

- *TFPI Kunitz Domain 2*: binds and inhibits FXa → docks FXa-TFPI to endothelium.
- *TFPI Kunitz Domain 1*: binds and inhibits TF-FVIIa → inactive TF-FVIIa/FXa-TFPI.

Importantly, TFPI cannot inhibit FVIIa until it is first bound to FXa, underlying its FXa-dependent activity. TFPI is specifically inhibiting the initiation stage of coagulation. If haemostasis is to proceed, the TFPI threshold must be breached, only possible through the recruitment of sufficient tissue factor, and thus, significant vessel wall damage [17, 18].

16.5.2 Activated Protein C

As illustrated earlier, as soon as FIXa is formed, a feedback cycle initiates that facilitates the formation of FXa and diminishes the efficacy of TFPI. As the coagulation cascade is now independent of TF, it can be regarded as in the 'propagation stage', with thrombin propagating haemostatic plug production. Activated protein C (APC) is instrumental in limiting the expansion of this plug to prevent thrombus expansion.

Protein C is a vitamin K-dependent serine protease zymogen synthesised in the liver, which circulates freely in the plasma and possesses the same conserved domain organisation as FVII, FIX and FX. On formation of a haemostatic plug, thrombin binds to thrombomodulin (TM) on the endothelial surface. Upon the binding of protein C to its receptor on the endothelium, the activation peptide is proteolytically removed by thrombin–thrombomodulin complexes, thereby activating it. The newly activated protein C can now cleave specific arginine residues to proteolytically inactivate FVa and FVIIIa, thereby removing both of the cofactors vital for the propagation of coagulation. This mechanism specifically target the propagation phase of coagulation, with no effect on the initiation phase.

Factor V Leiden is an autosomal dominant condition characterised by the mutation of a crucial arginine residue (Arg506) targeted by APC. With APC no longer able to efficiently inhibit factor V, its regulatory capacity is diminished, leading to the excessive production of thrombin and dramatically increasing the risk of thrombosis in affected patients.

The efficiency of activated protein C may be enhanced by its cofactor, Protein S. It is thought that Protein S not only enhances the affinity of activated protein C for phospholipid surfaces, but also relocates the active site of APC closer to the cleavage site in FVa and FVIIIa, enabling more efficient cleavage [19, 20].

16.5.3 Antithrombin

The final regulator of coagulation is antithrombin (AT): a serine protease inhibitor that circulates in the plasma at high concentrations to neutralise the risk of pathological thrombus formation posed by thrombin that may be washed downstream from the site of vascular injury. This inhibitory activity underlies the short half-life of soluble thrombin in plasma.

Despite its name, antithrombin inactivates several serine proteases, not just thrombin, with particular importance on those of the intrinsic pathway, namely FIXa, FXa, FXIa and FXIIa, in addition to FVIIa and thrombin.

Mechanistically, AT presents a 'bait loop' (mock substrate) to the serine protease. The interaction of this loop with the enzyme active site instigates a conformational change in AT that 'locks' the captured factor in an inactive 1:1 complex (thrombin:antithrombin). Fundamentally, this localises haemostasis to the site of injury. The importance of this is observable in people with antithrombin deficiency, who typically possess a significantly elevated risk of thrombus formation, pulmonary emboli and cerebrovascular incidents [21].

16.6 Clinical Implications: Anticoagulant Therapies

Therapeutic modulation of the coagulation cascade can modulate the risk of pathological thrombus formation. Key uses include acting as primary prophylaxis in atrial fibrillation, post-surgery and in genetic thrombophilia, in addition to using as secondary prophylaxis for conditions such as stroke, myocardial infarction and deep vein thrombosis [19].

16.6.1 Warfarin

Warfarin blocks the recycling of vitamin K by inhibiting vitamin K epoxide reductase and quinone reductase. Vitamin K is needed for the carboxylation and hence function of certain factors. By progressively depleting the supply of vitamin K, warfarin reduces the production of functional coagulation factors, specifically factors II, VII, IX and X. Over time, this will suppress thrombin (and hence fibrin) formation. Due to its indirect mechanism of action, warfarin exhibits slow pharmacodynamics, with the eventual exhaustion of coagulation factors dependent on their respective half-lives in circulation.

Warfarin is metabolised by CYP450, a liver enzyme whose activity is modulated by factors including diet, polymorphisms and other drugs. This variable bioavailability, coupled with a narrow therapeutic window, underscore the importance of continually monitoring the patient's coagulation status using INR (international normalised ratio), also referred to as *prothrombin time*.

To restore normal coagulation, warfarin should be stopped, and vitamin K administered. In an instance where immediate reversal is required, prothrombin complex concentrate (PCC), containing Factors II, VII, IX and X, can be administered [22, 23].

16.6.2 Heparin

If warfarin is contraindicated, for example in pregnancy, heparin is a common alternative. Heparin enhances the action of antithrombin by causing it to adopt its 'active' configuration. In this way, it amplifies the physiological anticoagulant pathways, reducing the activity of thrombin and factor Xa. Heparin comes in different forms: fondaparinux, unfractionated heparin and low molecular weight heparin (LMWH); and is administered via subcutaneous injection.

Fondaparinux comprises a synthetic version of the essential heparin pentasaccharide, incorporated into long polymers. In contrast, unfractionated heparin (UH) consists of larger molecules of less standard length, capable of bridging antithrombin molecules to stimulate anti-thrombin activity. LMWH mainly affects FXa inhibition and is characterised by a more uniform response deriving from its regular molecule length. Variation in molecule length is associated with a corresponding changeability in anticoagulant capacity. Because of this, heparin efficacy can be monitored by measuring the activated partial thromboplastin time (APTT). LMWH requires reduced monitoring due to a more uniform patient response. If required, protamine is able to bind and inhibit heparin, neutralising its effects. The indications for heparin are varied, however one key use of LMWH in particular is in the treatment of deep vein thrombosis (and potential pulmonary embolism) [24–26].

16.7 What We Don't Know: Usefulness of Novel Agents

Generally, the process of coagulation is a well understood topic. Growth in this field centres mainly around the identification of potential therapeutic targets and the subsequent development of safe, effective and modulatory therapies. The production of novel anticoagulants creates an exciting alternative to mainstay therapies such as warfarin, yet much of the current debate converges on the potential long-term utilisation of these neo-agents, with ongoing trials (both human and animal) evaluating their safety and efficacy [27].

16.8 Future Directions and Novel Therapies

16.8.1 Direct-Acting Oral Anticoagulants (DOACs)

The past 5 years has seen the gradual implementation of direct oral anticoagulants (DOACs) into clinical practice in an attempt to limit the use of warfarin and its variable pharmacokinetics. The therapeutic value of DOACs lies in their targetting of specific coagulation factors, rather than the generalised depletion of active factors achieved by vitamin K antagonists. This is exemplified by the thrombin-specific inhibition of dabigatran and targeting of FXa by rivaroxaban, edoxaban and apixaban [28].

Moreover, these small molecules have well-defined pharmacokinetics, with a fast onset and predictable activity, thus requiring less monitoring than their predecessors. The therapeutic value of DOACs is supported by an evidence base that generally demonstrates a reduced bleeding risk vs. vitamin K antagonists, with apixaban notably associated with a significant decrease in the incidence of intracranial bleeding compared to warfarin. It should be mentioned, however, that due to the exclusion of patients at an increased bleeding risk, further clinical investigations are needed before more firm conclusions can be reached [29].

In addition to a considerably higher price, an inherent limitation of DOACs is the difficulty in reversing their activity. At the time of writing, no established antidote for factor Xa inhibitors exists, with idarucizumab licensed in 2015 to reverse the effects of dabigatran. As a result, prothrombin complex concentrates (PCC) are usually given to replace lost factors in an attempt to restore coagulability [28, 29].

Take-Home Message

- Haemostasis is crucial to maintain the physiological balance between bleeding and thrombosis.
- Primary haemostasis is dependent on platelet activation and aggregation, via the action of vWF, ADP and TXA_2 and results in the formation of the primary haemostatic plug.
- Secondary haemostasis comprises of the coagulation cascade, resulting in amplification of thrombin production. The consequent formation and cross-linking of fibrin stabilises the platelet plug.
- Therapeutic manipulation of both primary and secondary haemostasis, as well as natural anticoagulant pathways, has revolutionised the management of prothrombic disorders, including myocardial infarction and cerebrovascular events.

References

1. Versteeg HH et al (2013) New fundamentals in hemostasis. Physiol Rev 93:327. https://www.physiology.org/doi/full/10.1152/physrev.00016.2011
2. Broos K et al (2011) Platelets at work in primary hemostasis. Blood Rev 25(4):155–167. https://doi.org/10.1016/j.blre.2011.03.002
3. Peyvandi F et al (2011) Role of von Willebrand factor in the haemostasis. Blood Transfus 9(Suppl 2):s3–s8. https://doi.org/10.2450/2011.002S

4. Gale AJ (2011) Current understanding of hemostasis. Toxicol Pathol 39(1):273–280. https://doi.org/10.1177/0192623310389474
5. Seong-Hoon Y et al (2016) Platelet activation: the mechanisms and potential biomarkers. Biomed Res Int 2016:1. https://doi.org/10.1155/2016/9060143
6. Blair P et al (2009) Platelet a-granules: basic biology and clinical correlates. Blood Rev 23(4):177–189. https://doi.org/10.1016/j.blre.2009.04.001
7. Hanby HA et al (2017) Platelet dense granules begin to selectively accumulate mepacrine during proplatelet formation. Blood Adv 1(19):1478–1490. https://doi.org/10.1182/bloodadvances.2017006726
8. Behan MWH, Storey RF (2004) Antiplatelet therapy in cardiovascular disease. Postgrad Med J 80:155–164. https://doi.org/10.1136/pgmj.2003
9. Bhatt DL, Topol EJ (2003) Scientific and therapeutic advances in antiplatelet therapy. Nat Rev Drug Discov 2(1):15–29. https://doi.org/10.1038/nrd985
10. Savi P et al (1998) Clopidogrel: a review of its mechanism of action. Platelets 9(3–4):251–255. https://doi.org/10.1080/09537109876799
11. Chatterjee S et al (2013) PAR-1 antagonists: current state of evidence. J Thromb Thrombolysis 35(1):1–9. https://doi.org/10.1007/s11239-012-0752-4
12. Horiuchi H (2006) Recent advance in antiplatelet therapy: the mechanisms, evidence and approach to problems. Ann Med 38(3):162–172. https://doi.org/10.1080/07853890600640657
13. Wang TH et al (2006) Aspiring and clopidogrel resistance: an emerging clinical entity. Eur Heart J 27(6):647–654. https://doi.org/10.1093/eurheartj/ehi684
14. Pozzi N et al (2013) Autoactivation of thrombin precursors. J Biol Chem 288(16):11601–11610. https://doi.org/10.1074/jbc.M113.451542
15. Lenting PJ et al (1998) The life cycle of coagulation factor VIII in view of its structure and function. Blood 92:3983–3996
16. Geddings JE (2014) New players in haemostasis and thrombosis. Thromb Haemost 111(4):570–574. https://doi.org/10.1160/TH13-10-0812
17. Crawley JTB, Lane DA (2008) The haemostatic role of tissue factor pathway inhibitor. Arterioscler Thromb Vasc Biol 28(2):233–242. https://doi.org/10.1161/ATVBAHA.107.141606
18. Wood et al (2014) Biology of tissue factor pathway inhibitor. Blood 123(19):2934–2943. https://doi.org/10.1182/blood-2013-11-512764
19. Castoldi E, Hackeng TM (2008) Regulation of coagulation by protein S. Curr Opin Haematol 15(5):529–536. https://doi.org/10.1097/MOH.0b013e328309ec97
20. Kalafatis M, Rand MD, Mann KG (1994) The mechanism of inactivation of human factor V and human factor Va by activated protein C. J Biol Chem 269(50):31869–31880
21. Rosenberg RD, Damus PS (1973) The purification and mechanism of action of human antithrombin-heparin cofactor. J Biol Chem 248(18):6490–6505
22. Lim GB (2015) Anticoagulants. Nat Rev Cardiol. https://doi.org/10.1038/nrcardio.2017.170
23. Hirsh J et al (2003) American Heart Association/American College of Cardiology Foundation guide to warfarin therapy. Circulation 107:1692–1711
24. Hirsh J et al (2001) Heparin and low-molecular-weight heparin mechanisms of action, pharmacokinetics, dosing, monitoring, efficacy and safety. Chest J 119(1):64S–94S. doi: 0.1378/chest.119.1_suppl.64S
25. Basu D et al (1972) A prospective study of the value of monitoring heparin treatment with the activated partial thromboplastin time. NEJM 287:324–327. https://doi.org/10.1056/NEJM197208172870703
26. Salzman EW et al (1975) Management of heparin therapy – controlled prospective trial. NEJM 292:1046–1050. https://doi.org/10.1056/NEJM197505152922002
27. Pollack CV Jr (2016) Coagulation assessment with the new generation of oral anticoagulants. BMJ 33(6):423–430
28. Wilson MR, Docherty KF, Gardner RS (2016) Use of direct oral anticoagulants in thromboembolic disease. Prescriber Drug Rev 27(8):15. https://doi.org/10.1002/psb.1487
29. Garcia D, Libby E, Crowther MA (2010) The new oral anticoagulants. Blood 115(1):15–20. https://doi.org/10.1182/blood-2009-09-241851

16

Cellular and Molecular Mechanisms of Atherosclerosis

Adam Tsao, Mridul Rana, and Joseph J. Boyle

© Springer Nature Switzerland AG 2019
C. Terracciano, S. Guymer (eds.), *Heart of the Matter*, Learning Materials in Biosciences,
https://doi.org/10.1007/978-3-030-24219-0_17

What You Will Learn in This Chapter

This chapter will comprehensively discuss the underlying pathophysiology of atherosclerosis: a disease of global importance, focusing particularly on the role of immune and smooth muscle cells. We will then explore current research surrounding factors that may contribute to plaque instability and subsequent haemorrhage, including novel theories regarding the potential role of various macrophage subsets. Finally, we will examine atherosclerosis at a signalling level, expanding upon the role of inflammatory recruitment pathways and its constituent molecules.

Learning Objectives

- Be able to discuss the primary pathophysiology of atherosclerosis and appreciate its chronic inflammatory profile.
- Explain the interplay of a variety of cell types, including macrophages and vascular smooth muscle cells (VSMCs).
- Discuss the role of Mhem macrophages in promoting atherosclerotic lesion stability.

17.1 The Global Atherosclerosis Burden

Atherosclerosis is a complex, chronic immunoinflammatory disease characterised by the progressive stenosis of blood vessels. This ancient pathology, identified in Egyptian mummies from 1580 BCE, has since become one of the world's foremost disease burdens, with the total direct and indirect costs predicted to exceed $1 trillion annually in the United States by 2035 [1].

Compounding this, the major clinical manifestations of atherosclerosis, namely myocardial infarction (MI), ischaemic heart disease (IHD) and peripheral artery disease (PAD), constitute a considerable morbidity and mortality burden, particularly in developing nations [2]. The magnitude of this disease has driven the need to greater understand its complex pathophysiology, with risk factors such as age, male sex, family history, high blood pressure, smoking, and diabetes all contributing to atherosclerotic plaque development [3].

Considerable evidence also highlights the increasingly integral role of inflammation and the multifaceted interplay this has with the aforementioned 'traditional' risk factors [2, 3]. As the global population continues to age, the treatment of atherosclerosis and its associated co-morbidities will remain of global importance.

17.2 Endothelial Dysfunction

17.2.1 Nitric Oxide

17

Prior to the formation of an atheroma, dysregulation and dysfunction of the endothelium represents a critical yet reversible deviation in its role of orchestrating vascular physiology [4]. Arguably the foremost example of this is the endothelium-dependent secretion of nitric oxide (NO): a potent anti-atherogenic by-product of the nitric oxide synthase III (eNOS)-catalysed conversion of L-arginine to L-citrulline [5]. NO diffuses to vascular smooth muscle cells (VSMCs) in the underlying tunica media, stimulating relaxation and thus vasodilation via the activation of guanylate cyclase and intracellular cyclic guanosine monophosphate (cGMP) [5].

In acute inflammatory situations, e.g., infection, cytokine-mediated activation of endothelial cells (ECs) induces expression of adhesion molecules including VCAM-1, ICAM-1 and E-selectin, facilitating the localised recruitment of immune cells. The capacity of NO to downregulate processes highlights its homeostatic and anti-inflammatory importance, exemplified by the aberrant expression of adhesion molecules in pathological endothelial dysfunction [6]. Mechanistically, NO exerts this downregulation via the stabilisation of IκβαΙ, an endogenous inhibitor of NF-κβ transcription factors, which interacts with a common DNA binding motif to facilitate the expression of pro-inflammatory mediators like adhesion molecules [7]. Other vasculoprotective effects of NO include its role in inhibiting VSMC proliferation and downregulating platelet aggregation, both of which are anti-atherogenic [8].

With this in mind, it is not surprising that diminished NO bioavailability, a defining feature of endothelial dysfunction, is so widely considered a critical step in the pathogenesis of atherosclerosis [6].

Endothelial dysfunction (ED) represents a convergent output of atherogenic risk factors including chronic inflammation, hyperglycaemia and smoking. These noxious stimuli are all associated with eNOS uncoupling, resulting in the production of reactive oxygen species (ROS) that go on to oxidise BH_4, inhibiting its co-factor role in NO synthesis [9]. Furthermore, ROS also inactivates existing NO, accelerating its diminishment [10].

The consequential NO deficit indirectly promotes the expression of pro-adhesion molecules, increasing pro-inflammatory signalling and dysregulating vascular tone. Lacking NO inhibition, ROS induces a state of 'oxidative stress' that mediates the modification of subendothelial low-density lipoproteins (LDLs) into an oxidised form (oxLDLs) [11]. This is summarised in ◘ Fig. 17.1.

17.2.2 **Atheroprone Arterial Regions**

Classical cardiovascular (CV) risk factors, e.g., diabetes, smoking, hyperlipidaemia, are systemic, yet atherosclerosis is spatially nonrandom, arising at predictable 'atheroprone' focal points in the vasculature, namely curvatures and branch points [12]. The current consensus is that ECs transduce haemodynamic forces into signalling events via mechanosensors. Indeed, atheroprone regions demonstrate a nonlaminar flow with reduced shear stress: the mechanical, frictional force exerted by blood flow onto the endothelium [12]. High velocity laminar flow transmits high shear to the endothelium, associated with

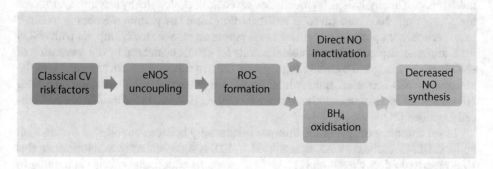

◘ **Fig. 17.1** Schematic showing the interplay of classical CV risk factors with endothelial dysfunction

relative atheroprotection: a relationship demonstrated both in animal models, and by spatially mapping human plaques [13, 14].

Existing literature supports the phenomenon of high shear stress stimulating Ca^{2+}-activated K^+ channels, hyperpolarising the endothelium and increasing the driving force of Ca^{2+} ions into the endothelium [15]. This amplifies eNOS activity, increasing endothelium-dependent NO synthesis. Conversely, Dimmeler et al. ascribes the upregulated NO production to the activation of integrins and surface complexes such as the glycocalyx, which facilitate the phosphorylation and subsequent enhancement of eNOS activity via Akt/PKB signalling [16].

Both proposals have valid datasets, with multiple mechanotransduction mechanisms likely interacting to cumulatively upregulate NO synthesis.

Accordingly, the decreased shear stress exerted by turbulent or oscillatory flow precipitates constitutively lower NO production and thus pro-inflammatory signalling [16]. This is strongly linked with degradation of the glycocalyx: a glycolipid-glycoprotein network that covers the endothelium, regulating intimal permeability. Indeed, ROS can facilitate the breakdown of the glycocalyx both directly and via activation of proteoglycan-targeting matrix metalloproteinases, namely MMPs-2, -7 and -9 [17]. This interplay is reaffirmed by the structurally-compromised glycocalyx and upregulated ROS observed in pro-inflammatory states, especially diabetes [18].

Finally, nonlaminar flow has been shown to upregulate VCAM-1 expression [19]. With increased endothelial permeability already facilitating LDL entry, upregulation of VCAM-1 and other pro-inflammatory signals catalyses the adhesion and transmigration of inflammatory cells, promoting both ROS production and the oxidative modification of LDLs to oxLDLs (oxidised LDL cholesterols). This is supported by the downregulation of endothelium-dependent NO synthesis, which would otherwise inhibit VCAM-1 and inflammatory signalling via Iκβα [7]. Over time, this progressive escalation of subendothelial oxLDL and inflammatory cell accumulation promotes fatty streak formation: a critical morphological indicator of early atherosclerosis.

17.3 Oxidised LDLs

Upregulation of ROS production progressively overwhelms endogenous antioxidant mechanisms, facilitating the modification of LDL to oxLDL: a highly immunogenic compound that mediates fatty streak formation and further supports the inflammatory-mediated depletion of NO. OxLDL inhibits EC-dependent NO production via two main mechanisms. Chiefly, it depletes membrane caveolae of cholesterol by serving as an acceptor, promoting eNOS and caveolin redistribution from the plasma membrane to inner sites. Secondly, it upregulates arginase I expression: an enzyme that competes with eNOS for L-arginine, depleting the available substrate for eNOS, evidenced by the restoration of eNOS activity instigated by inhibiting arginase [20]. In this way, oxLDL contributes to the removal of NO-mediated Iκβα inhibition of pro-atherogenic NF-κβ signalling, with NF-κβ further increasing endothelial permeability via IL-8-mediated modification of cell-cell junctions [21].

In yet another example of the intimate relationship between endothelial dysfunction and oxLDL, thromboxane A2, upregulated in ED, is independently able to initiate this IL-8-mediated effect via NF-κβ [21]. OxLDL exerts its endothelial effects via binding to lectin-like oxidised LDL receptor-1 (LOX-1), encoded by the OLR1 gene [22]. OLR1 expression is characteristically upregulated in ED, driven by a pro-inflammatory milieu including oxLDL, IL-6 and TNF-α, all of which exert this effect [22, 23].

17.3.1 Monocyte Recruitment

Prior to morphological endothelial damage, a significant monocyte population is already present in the subintimal space of atheroprone vasculature in early endothelial dysfunction, supporting the notion that it is a compensatory response to oxidative stress induced by aberrant signalling and nonlaminar flow.

The past three decades have seen the identification of signalling pathways behind the upregulation of cell adhesion molecules (CAMs) responsible for the atherogenic recruitment of monocytes from the blood, including VCAM-1, ICAM-1, E-selectin and P-selectin [24, 25]. These mechanisms all include an underlying Ox-LDL/LOX-1/CD40/PKC interaction, the importance of which was confirmed by Li et al. in three key findings [26]:

1. LOX-1 inhibition significantly reduced oxLDL-mediated CD40 expression.
2. CD40 inhibition significantly reduced oxLDL-mediated TNF-α release and CAM expression.
3. PKC inhibition markedly reduced oxLDL-mediated CD40 expression with LOX-1 antibodies blocking PKC activation.

Monocyte recruitment to the atherogenic site involves endothelial adhesion and transmigration, facilitated by CAMs and the expression of chemokines such as MCP-1, which are also upregulated by Ox-LDL/LOX-1/CD40/PKC signalling [27].

17.4 Macrophage Activation

Considerable evidence suggests that the dynamic differentiation process of subendothelial monocytes into macrophages (MØ) is initially dependent on colony-stimulating cytokines, especially M-CSF and GM-CSF, with both being upregulated by oxLDL-induced signalling pathways [28, 29]. After this common initiatory step, a complex signalling nexus determines the macrophage subsets, the two foremost of which are M1 and M2 [30]. These distinctions are highlighted in ◘ Fig. 17.2.

Both subsets have been identified in endarterectomised human plaques, with greater M1 representation than M2 [31]. This finding in mature lesions is complementary to the subset transition hypothesis proposed by Khallou-Laschet et al., who demonstrated that murine lesion progression correlated with a significant phenotypic shift from M2 to M1 over time [32]. To exclude the possibility that this was attributable to augmented M1 plaque infiltration, further repolarisation experiments provided sufficient evidence that this shift was a phenotypic conversion of already-present cells [32].

This was furthered by van Tits et al., who detected this shift following abundant M2 uptake of oxLDL, mediated by scavenger receptors, especially CD36 [33]. Interestingly, Krüppel-like factor 2 (KLF-2), an anti-inflammatory transcription factor native to monocytes and ECs, decreased following oxLDL uptake, promoting phenotypic shift to pro-inflammatory M1-like cells, with upregulated production of inflammatory cytokines IL-6, IL-8 and MCP-1 [33]. OxLDL exposure to M1 MØ induced negligible KLF-2 and phenotypic changes, whilst subsequent siRNA-downregulation of KLF-2 expression in M2 reproduced the aforementioned MCP-1 upregulation with high fidelity.

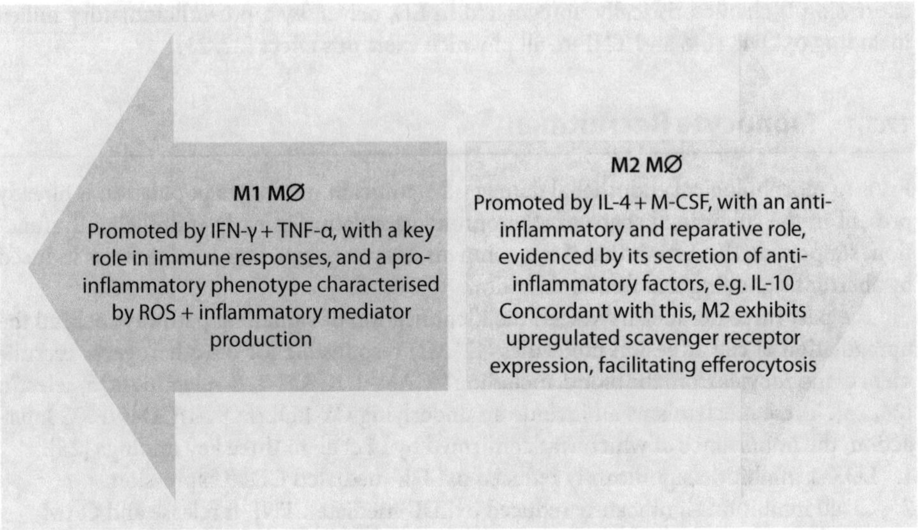

Fig. 17.2 Overview of the two key MØ subsets in atherosclerotic signalling

This suggests that scavenger receptor-mediated uptake of oxLDL acts via KLF-2 down-regulation to change the MØ inflammatory profile, with the relative overexpression of scavenger receptors in M2 precipitating its apparent susceptibility to foam cell formation. With KLF-2 known to downregulate NF-κβ signalling, this proposed pathway is compatible with the NF-κβ upregulation observed upon M2 MØ oxLDL phagocytosis [33, 34].

In an elegant bidirectional relationship, oxLDL serves to enhance CD36 expression and pro-inflammatory signalling in M2, augmenting its atherogenic uptake and expanding M2 representation in early fatty streaks. Reciprocally, this now M1-like population has greater capacity to modify LDL to immunogenic forms, which are then phagocytosed, perpetuating both these relationships in a synergistic manner to advance atherogenesis.

17.5 Atherosclerotic Lesions

17.5.1 Fatty Streak Formation

The modification of LDL to oxLDL, including the oxidation of apolipoprotein B (ApoB) and subsequent amino acid side-chain alterations, facilitates uptake by macrophage scavenger receptors [35]. Two key interactions are those of ox-ApoB and class A scavenger receptors, and oxidised phospholipids with class B scavenger receptors (e.g., CD36), with the latter of greater importance in triggering oxLDL internalisation and subsequent macrophage activation. Following internalisation, oxLDL lipids are able to activate PPAR-γ transcription factors, upregulating CD36 expression and thus oxLDL uptake. Macrophage activation also upregulates ROS synthesis, catalysing the oxidative modification of further LDLs.

Moreover, following oxLDL internalization, ACAT1 and nCEH enzymes facilitate esterification and further hydrolysis, eventually producing cytoplasmic lipid droplets. Whilst this uptake is normally balanced by cholesterol efflux mediated chiefly by ABCA1, several critical efflux transporters have been shown to be downregulated by oxLDL-induced signalling events in both macrophages and VSMCs [36, 37]. This, coupled with upregulated oxLDL internalization, overwhelms cholesterol homeostasis, with consequential lipid droplet accumulation and foam cell formation over time.

This paradigm would suggest that lipid accumulation may be retarded by the reinstitution of cholesterol homeostasis, and indeed, Yoshinaka et al. demonstrated that selective inhibition of ACAT-1 does prevent foam cell formation in ApoE$^{-/-}$ mice [37].

17.5.2   Plaque Maturation

Over time, progressive dysregulation of intracellular cholesterol and the consequential escalating ER stress triggers unfolded protein response (UPR) in the ER, leading to apoptosis of VSMC or MØ foam cells via both the mitochondrial and Fas pathways. In early lesions, this process is non-inflammatory, with phosphatidylserine on the membranes of apoptotic bodies interacting with scavenger receptors, facilitating efferocytosis [38]. With lesion advancement, this anti-inflammatory mechanism ultimately fails for three key reasons:

1. Foam cells upregulate both oxLDL production and internalisation → more foam cells → progressive increase in apoptosis.
2. Foam cells represent an increased proportion of total MØ over time → reduced efferocytosis capacity for apoptotic bodies.
3. Intraplaque haemorrhage more frequent in advanced lesions → phagocytosis of erythrocytes hinders MØ phagocytosis capacity [38].

The progressive insufficiency of scavenger-mediated efferocytosis subsequently results in an autolytic, necrotic process termed *secondary necrosis* [39]. Over time, the intraplaque accumulation of necrotic debris, extracellular free cholesterol and pro-apoptotic cytokines triggers further apoptosis in nearby cells, creating an intensely inflammatory necrotic core [38].

17.6   Plaque Destabilisation

Amongst the panoply of inflammatory mediators released during foam cell formation are MCP-1, promoting VSMC proliferation and leukocyte migration, and TNF-α [40]. By binding to ECs, TNF-α uses NF-κβ signalling to enhance existing endothelial dysfunction and promote further CAM upregulation, which manifests as escalating recruitment of VSMCs and immune cells.

The proliferating intraplaque VSMC population transiently stabilises the developing lesion via synthesis of collagen and other ECM proteins that sustain structural integrity of the fibrous cap. Following the observation that high macrophage density was associated with a weaker fibrous cap and lesional instability, Libby et al. provided the eventual explanation that metalloproteinase production by the expanded population of pro-inflammatory macrophages was responsible for cap degradation and thus the enhanced rupture risk [27]. MMP-mediated ECM degradation, coupled with incremental VSMC apoptosis progressively heightens the risk of lesional rupture, exposing subendothelial thrombogenic compounds, e.g., tissue factor, causing platelet aggregation, thrombus formation and vascular occlusion.

17.7 What We Don't Know: The Changing MØ Paradigm

The association between intraplaque haemorrhage and lesion progression is well established, with the cholesterol-rich membranes of erythrocytes dramatically increasing the total free cholesterol content in the necrotic core, accelerating both lesional enlargement and destabilisation. Moreover, erythrocyte-derived haemoglobin provides a source of iron, catalysing the conversion of hydrogen peroxide to highly ROS, triggering additional oxidative damage and furthering macrophage recruitment [41].

Traditionally, macrophages have been considered to take on either pro-inflammatory or anti-inflammatory phenotypes depending on their environment. As previously discussed, these were termed M1 and M2, with M1 characterised by high levels of atherogenic IL-6 and INF-γ, and the latter exhibiting anti-inflammatory activity through the secretion of IL-10 and TGF-β. However, this has been complicated by the identification and characterisation of additional macrophage phenotypes, including M4, Mhem, Mox, rM and Mreg.

For example, Boyle et al. identified an erythrophagocytic 'Mhem' MØ subset. This is characterised by upregulated CD163 expression, facilitating scavenging of haemoglobin-haptoglobin complexes with subsequent secretion of anti-inflammatory IL-10, in addition to low human leukocyte antigen-DR (HLA-DR) [42]. Interestingly, IL-10 secretion is also a salient feature of M2 macrophages, prior to the aforementioned oxLDL-KLF-2-mediated shift in signalling profile [42].

Unlike M2, Mhem macrophages do not exhibit foam cell formation and retain IL-10 secretion, reaffirming their atheroprotective role. In vitro studies demonstrate that haem is a crucial stimulus for the IL-10-dependent Mhem differentiation process from monocytes, suggesting a role in intraplaque haemorrhage [42]. Mechanistically, haem appears to regulate the expression of activating transcription factor 1 (ATF1), whose subsequent activation via phosphorylation allows the co-regulation of HO-1 and LXRβ expression [43].

- HO-1 catalyses the conversion of haem → biliverdin + free iron + carbon monoxide. Biliverdin is thereafter converted to antioxidant bilirubin, whilst low concentration of carbon monoxide triggers guanylate cyclase-mediated atheroprotective signalling. HO-1 also upregulates ferritin, chelating the free iron [43].
- LXRβ upregulates both ABCA1 and ABCG1, utilising LXRα to augment cholesterol efflux via HDL synthesis [43].

The concurrent activity of these mechanisms facilitates the parallel processing of erythrocyte iron and cholesterol, neutralising their mutually pro-inflammatory and respective pro-oxidant and pro-foam cell effects. It is important to mention, however, that integration of these additional sub-types into the existing dichotomous framework has proven challenging, and realistically it is likely that macrophage phenotypes manifest as a spectrum between pro- and-anti-inflammatory activity. Other immune cells have also been identified as important in lesion progression. The role of neutrophils appears to be important; whilst dendritic cells, T cells and B cells have all been suggested as having meaningful roles in the MØ response to plaque signalling [44] (◘ Fig. 17.3).

17

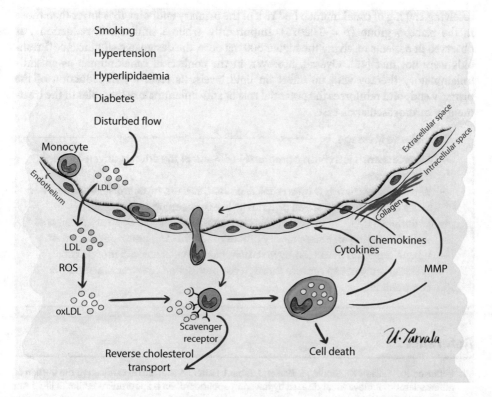

Smoking

Hypertension

Hyperlipidaemia

Diabetes

Disturbed flow

Monocyte

Extracellular space

Intracellular space

Endothelium

LDL

LDL

ROS

oxLDL

Collagen

Chemokines

Cytokines

MMP

Scavenger
receptor

Reverse cholesterol
transport

Cell death

U. Turvala

▫ Fig. 17.3 A summary of atherosclerotic lesion development

17.8 Where We're Heading: Focussing on Inflammation

For patients with pre-existing atherosclerotic plaques, diagnostic and therapeutic approaches are likely to change over the coming years. As plaque stability does not necessarily correlate with plaque size, it is difficult to predict which lesions are at risk of rupture or erosion. Developing a technique to identify vulnerable plaques, perhaps based upon inflammatory activity, would be invaluable in the risk stratification of patients. Moreover, due to the immunological nature of atherosclerosis, the future may see the development of anti-inflammatory or pro-resolution treatments to decelerate plaque progression, exemplified by the increasingly popular concept of a vaccine to develop adaptive immunity to the pathogenic inflammatory processes underlying atherosclerosis.

A key study to consider is the 2017 CANTOS (Canakinumab Anti-inflammatory Thrombosis Outcome Study) trial, which investigated the effects of canakinumab, a human monoclonal antibody that targets interleukin (IL)-1β: a cytokine in the IL-6 inflammatory signalling pathway. In this double-blind trial, 10,061 patients with previous MI were randomised to either one of three canakinumab doses (50 mg, 150 mg, 300 mg) or placebo [45]. It is important to note that none of the three doses had any notable effect on lipid levels.

Interestingly, only the 150 mg dose was associated with a statistically significant effect on the measured primary endpoints of nonfatal MI, nonfatal stroke and CV death. Patients receiving 150 mg of canakinumab had risk of the primary endpoint 15% lower than those in the placebo group ($p = 0.02075$). Importantly, while a similar risk reduction was observed in patients receiving the higher 300 mg dose, the designated significance thresholds were not met [45]. Overall, however, in the context of canakinumab as an anti-inflammatory therapy with no effect on lipid levels, its relative risk reduction of the primary endpoint reinforces the potential role of anti-inflammatory therapies in the treatment of cardiovascular disease.

> **Take-Home Message**
> - Atherosclerosis is a chronic progressive disease of the arteries driven by sterile inflammation.
> - While atherosclerosis is often clinically silent, strokes, myocardial infarctions and acute limb ischaemia are all potentially life-threatening complications.
> - Endothelial dysfunction allows low-density lipoproteins to deposit into arterial walls, triggering an inflammatory process and initiating lesion development.
> - Plaque growth is advanced by macrophages; vascular smooth muscle cells (VSMCs) prevent cap rupture through the production of extracellular matrix components.

References

1. Robinson JG, Williams KJ, Gidding S, Boršen J, Tabas I, Fisher EA et al (2018) Eradicating the burden of atherosclerotic cardiovascular disease by lowering apolipoprotein b lipoproteins earlier in life. J Am Heart Assoc 7:e009778
2. Celermajer DS, Chow CK, Marijon E, Anstey NM, Woo KS (2012) Cardiovascular disease in the developing world: Prevalences, patterns, and the potential of early disease detection. J Am Coll Cardiol 60:1207
3. Smith SC (2007) Multiple risk factors for cardiovascular disease and diabetes mellitus. Am J Med 120:S3
4. Anderson TJ, Meredith IT, Yeung AC, Frei B, Selwyn AP, Ganz P (1995) The effect of cholesterol-lowering and antioxidant therapy on endothelium-dependent coronary vasomotion. N Engl J Med 332(8):488–493
5. Förstermann U, Sessa WC (2012) Nitric oxide synthases: regulation and function. Eur Heart J 33:829
6. Deanfield JE, Halcox JP, Rabelink TJ (2007) Endothelial function and dysfunction. Circulation 115(10):1285–1295
7. Peng HB, Libby P, Liao JK (1995) Induction and stabilization of IκBα by nitric oxide mediates inhibition of NF-κB. J Biol Chem 270(23):14214–14219
8. Garg UC, Hassid A (1989) Nitric oxide-generating vasodilators and 8-bromo-cyclic guanosine monophosphate inhibit mitogenesis and proliferation of cultured rat vascular smooth muscle cells. J Clin Invest 83(5):1774–1777
9. Landmesser U, Dikalov S, Price SR, McCann L, Fukai T, Holland SM et al (2003) Oxidation of tetrahydrobiopterin leads to uncoupling of endothelial cell nitric oxide synthase in hypertension. J Clin Invest 111(8):1201–1209
10. Rubanyi GM, Vanhoutte PM (1986) Superoxide anions and hyperoxia inactivate endothelium-derived relaxing factor. Am J Phys 250(5 (Part 2)):H822–H827
11. Cominacini L, Garbin U, Pasini AF, Davoli A, Campagnola M, Pastorino AM et al (1998) Oxidized low-density lipoprotein increases the production of intracellular reactive oxygen species in endothelial cells: inhibitory effect of lacidipine. J Hypertens 16(12 II):1913–1919
12. Ku DN, Giddens DP, Zarins CK, Glagov S (1985) Pulsatile flow and atherosclerosis in the human carotid bifurcation. Positive correlation between plaque location and low oscillating shear stress. Arterioscler Thromb Vasc Biol 5(3):293–302

13. Winkel LC, Hoogendoorn A, Xing R, Wentzel JJ, Van der Heiden K (2015) Animal models of surgically manipulated flow velocities to study shear stress-induced atherosclerosis. Atherosclerosis 241:100–110

14. Zarins CK, Giddens DP, Bharadvaj BK, Sottiurai VS, Mabon RF, Glagov S (1983) Carotid bifurcation atherosclerosis. Circ Res 53(4):502–515

15. Brakemeier S, Kersten A, Eichler I, Grgic I, Zakrzewicz A, Hopp H et al (2003) Shear stress-induced up-regulation of the intermediate-conductance Ca2+–activated K+channel in human endothelium. Cardiovasc Res 60(3):488–496

16. Dimmeler S, Fleming I, Fisslthaler B, Hermann C, Busse R, Zeiher AM (1999) Activation of nitric oxide synthase in endothelial cells by Akt-dependent phosphorylation. Nature 399(6736):601–605

17. Lipowsky HH, Lescanic A (2013) The effect of doxycycline on shedding of the glycocalyx due to reactive oxygen species. Microvasc Res 90:80–85

18. Kolářová H, Ambrůzová B, Švihálková Šindlerová L, Klinke A, Kubala L (2014) Modulation of endothelial glycocalyx structure under inflammatory conditions. Mediat Inflamm 2014:1

19. Son DJ, Kumar S, Takabe W, Woo Kim C, Ni CW, Alberts-Grill N et al (2013) The atypical mechanosensitive microRNA-712 derived from pre-ribosomal RNA induces endothelial inflammation and atherosclerosis. Nat Commun 4:3000

20. Wang W, Hein TW, Zhang C, Zawieja DC, Liao JC, Kuo L (2011) Oxidized low-density lipoprotein inhibits nitric oxide-mediated coronary arteriolar dilation by up-regulating endothelial arginase I. Microcirculation 18(1):36–45

21. Kim S-R, Bae S-K, Park H-J, Kim M-K, Kim K, Park S-Y et al (2010) Thromboxane A(2) increases endothelial permeability through upregulation of interleukin-8. Biochem Biophys Res Commun 397(3):413–419

22. Tatsuguchi M, Furutani M, Hinagata J, Tanaka T, Furutani Y, Imamura S et al (2003) Oxidized LDL receptor gene (OLR1) is associated with the risk of myocardial infarction. Biochem Biophys Res Commun 303(1):247–250

23. Chen M, Kakutani M, Minami M, Kataoka H, Kume N, Narumiya S et al (2000) Increased expression of lectinlike oxidized low density lipoprotein receptor-1 in initial atherosclerotic lesions of Watanabe heritable hyperlipidemic rabbits. Arterioscler Thromb Vasc Biol 20(4):1107–1115

24. Yellin MJ, Brett J, Baum D, Matsushima A, Szabolcs M, Stern D et al (1995) Functional interactions of T cells with endothelial cells: the role of CD40L-CD40-mediated signals. J Exp Med 182(6):1857–1864

25. Alderson MR (1993) CD40 expression by human monocytes: regulation by cytokines and activation of monocytes by the ligand for CD40. J Exp Med 178(2):669–674

26. Li D, Liu L, Chen H, Sawamura T, Mehta JL (2003) LOX-1, an oxidized LDL endothelial receptor, induces CD40/CD40L signaling in human coronary artery endothelial cells. Arterioscler Thromb Vasc Biol 23(5):816–821

27. Galis ZS, Sukhova GK, Lark MW, Libby P (1994) Increased expression of matrix metalloproteinases and matrix degrading activity in vulnerable regions of human atherosclerotic plaques. J Clin Invest 94(6):2493–2503

28. Seo JW, Yang EJ, Yoo KH, Choi IH (2015) Macrophage differentiation from monocytes is influenced by the lipid oxidation degree of low density lipoprotein. Mediat Inflamm 2015:1

29. Kiyan Y, Tkachuk S, Hilfiker-Kleiner D, Haller H, Fuhrman B, Dumler I (2014) OxLDL induces inflammatory responses in vascular smooth muscle cells via urokinase receptor association with CD36 and TLR4. J Mol Cell Cardiol 66:72–82

30. Martinez FO, Gordon S (2014) The M1 and M2 paradigm of macrophage activation: time for reassessment. F1000Prime Rep 6:6–13

31. Shaikh S, Brittenden J, Lahiri R, Brown PAJ, Thies F, Wilson HM (2012) Macrophage subtypes in symptomatic carotid artery and femoral artery plaques. Eur J Vasc Endovasc Surg 44(5):491–497

32. Khallou-Laschet J, Varthaman A, Fornasa G, Compain C, Gaston AT, Clement M et al (2010) Macrophage plasticity in experimental atherosclerosis. PLoS One 5(1):e8852

33. Van Tits LJH, Stienstra R, van Lent PL, Netea MG, Joosten LAB, Stalenhoef AFH (2011) Oxidized LDL enhances pro-inflammatory responses of alternatively activated M2 macrophages: a crucial role for Krüppel-like factor 2. Atherosclerosis 214(2):345–349

34. Groeneweg M, Kanters E, Vergouwe MN, Duerink H, Kraal G, Hofker MH et al (2006) Lipopolysaccharide-induced gene expression in murine macrophages is enhanced by prior exposure to oxLDL. J Lipid Res 47(10):2259–2267

35. Henriksen T, Mahoney EM, Steinberg D (1981) Enhanced macrophage degradation of low density lipoprotemi previously incubated with cultured endothelial cells: recognition by receptors for acetylated low density lipoproteins. Med Sci 78(10):6499–6503

36. Guo L, Chen CH, Zhang LL, Cao XJ, Ma QL, Deng P et al (2015) IRAK1 mediates TLR4-induced ABCA1 downregulation and lipid accumulation in VSMCs. Cell Death Dis 6(10):e1949

37. Yoshinaka Y, Shibata H, Kobayashi H, Kuriyama H, Shibuya K, Tanabe S et al (2010) A selective ACAT-1 inhibitor, K-604, stimulates collagen production in cultured smooth muscle cells and alters plaque phenotype in apolipoprotein E-knockout mice. Atherosclerosis 213(1):85–91

38. Kockx MM, Herman AG (2000) Apoptosis in atherosclerosis: beneficial or detrimental? Cardiovasc Res 45:736–746

39. Wyllie AH, Kerr JFR, Currie AR (1980) Cell death: the significance of apoptosis. Int Rev Cytol 68(C): 251–306

40. Chen C, Khismatullin DB (2015) Oxidized low-density lipoprotein contributes to atherogenesis via co-activation of macrophages and mast cells. PLoS One 10(3): e0123088

41. Kolodgie FD, Gold HK, Burke AP, Fowler DR, Kruth HS, Weber DK et al (2003) Intraplaque hemorrhage and progression of coronary atheroma. N Engl J Med 349(24):2316–2325

42. Boyle JJ, Harrington HA, Piper E, Elderfield K, Stark J, Landis RC et al (2009) Coronary intraplaque hemorrhage evokes a novel atheroprotective ma0crophage phenotype. Am J Pathol 174(3):1097–1108

43. Boyle JJ, Johns M, Lo J, Chiodini A, Ambrose N, Evans PC et al (2011) Heme induces heme oxygenase 1 via Nrf2: role in the homeostatic macrophage response to intraplaque hemorrhage. Arterioscler Thromb Vasc Biol 31:2685–2691

44. Libby P (2002 Dec 19) Inflammation in atherosclerosis. Nature 420(6917):868–874

45. Ridker PM, Everett BM, Thuren T, MacFadyen JG, Chang WH, Ballantyne C et al (2017) Antiinflammatory therapy with canakinumab for atherosclerotic disease. N Engl J Med 377:1119

17

Molecular and Cellular Mechanisms of Angiogenesis

Zarius Ferozepurwalla, Jude Merzah, Lieze Thielemans, and Graeme Birdsey

© Springer Nature Switzerland AG 2019
C. Terracciano, S. Guymer (eds.), *Heart of the Matter*, Learning Materials in Biosciences,
https://doi.org/10.1007/978-3-030-24219-0_18

What You Will Learn in This Chapter

This chapter provides an overview of sprouting angiogenesis. We will begin by discussing the importance of vascular endothelial growth factor (VEGF) as the molecular conductor of this process, including its receptor interactions and subsequent signalling. At each step of this process, both the observable cellular changes and underlying molecular pathways will be explored. Competing models of lumen formation will be critically appraised, before we examine the reestablishment of vascular quiescence and importance of junctional stability via VE-cadherin interactions.

Learning Objectives

- Understand the stages of sprouting angiogenesis in a systematic order, including both observable cellular changes and the molecular signalling underlying them.
- Explain the importance and mechanism of Dll4-Notch signalling in the activation of the endothelium and characterisation of tip and stalk cell phenotypes.

18.1   Introduction to Angiogenesis

Living organisms require an extensive vascular network for the provision of adequate nutrients and oxygen commensurate to metabolic demand, in addition to the removal of waste products and toxins from local tissue areas. As organisms have grown in size and complexity, an increasingly sophisticated apparatus is required. Notably, this includes the integral capacity to create and add to an existing vascular network in order to meet a changeable metabolic demand. Three main mechanisms of expanding the vasculature exist [1]:

1. *Vasculogenesis*: the de novo differentiation of vascular endothelial cells from their angioblast progenitors.
2. *Arteriogenesis*: the remodelling of existing collateral vessels to enhance patency and relieve ischaemia.
3. *Angiogenesis*: the formation of neo-vessels from pre-existing blood vessels [2].

This chapter will focus on angiogenesis, which is ultimately responsible for the majority of new vessel growth, both during development and in disease. Though primarily switched off, angiogenesis is vital in processes such as wound healing and menstruation, as well as during embryonic vascular development [2].

18.2   Vascular Endothelial Growth Factor

An interacting network of stimulatory and inhibitory signalling molecules closely regulates angiogenesis, with the foremost stimulatory factor, vascular endothelial growth factor (VEGF), having a vital and well-established role, as evidenced by the embryonic lethality seen following deletion of one or both of its encoding alleles [3, 4]. After binding to its receptor, the vascular endothelial growth factor receptor (VEGFR), a DII4/Notch signalling cascade results in myriad expression changes that ultimately initiate the process of angiogenesis [3, 5]. This is observable as a distinct series of phenotypic changes in the endothelium that will be discussed later.

Counteracting this is a series of inhibitory factors such as angiostatin and endostatin, which suppress angiogenesis via mechanisms that include the inhibition of VEGF [6, 7]. The product of this overall signalling nexus is a highly regulated process, which, aided by transcription factors (TFs), controls the expression of many endothelial genes in angiogenesis.

18.2.1 Initiation

Hypoxia is the initial stimulus for angiogenesis. Cellular oxygen homeostasis is dependent upon the activity of *hypoxia-inducible factors* (*HIFs*): a heterodimeric transcription factor family that functions as a vital component in the cellular hypoxia response via its O_2-sensitive α-subunit [8].

In environments of adequate oxygen (normoxia), HIFs are hydroxylated by oxygen-sensing *prolyl hydroxylase domain enzymes* (*PHD*), permitting the ubiquitination of HIFs to mark them for degradation by proteasomes [8]. There are two key processes by which this occurs:

1. Proline residues on HIF undergo trans-4-hydroxylation, promoting interactions with the von Hippel-Lindau complex (VHL) to target HIFs for proteasomes [8].
2. Asparagine residues are β-hydroxylated, downregulating HIF transactivation [9].

In the absence of O_2, the inactivity of PHD results in the retainment of HIF activation, thereby allowing HIFs to translocate to the nucleus and upregulate the transcription of angiogenic factors, the most important of which is VEGF-A [2]. The release of pro-angiogenic factors from stromal cells acts as the initiator of the angiogenic process, increasing the oxygen supply to surrounding tissues in an attempt to reverse the hypoxia [2].

18.2.2 Angiogenic Inductive Signalling

In the healthy adult, the lumen of mature blood vessels is encircled by a monolayer of endothelial phalanx cells. Control of endothelial homeostasis is via cell-cell junctions, which utilise junctional molecules such as VE-cadherin and claudins to regulate both the survival and integrity of the monolayer [10]. Structurally, the homophilic interactions of VE-cadherin on adjacent endothelial cells catalyses the tight binding of cells into a continuous endothelial architecture.

Enveloping this endothelial tube are mural cells such as vascular smooth muscle cells (VSMCs) and pericytes. These serve to dampen endothelial proliferation and release stability signals including angiopoietin-1 (ANG-1) in order to promote a quiescent vascular phenotype [2, 6].

Upon recognition of pro-angiogenic stimuli, namely VEGF, angiopoietin-2 (ANG-2), fibroblast growth factor-2 (FGF-2), or chemokines, pericytes detach from the vascular wall and VEGF commences angiogenesis through activation of the endothelium and subsequent changes in endothelial gene expression and phenotype [2, 6].

To appreciate sprouting angiogenesis at the cellular level, it is vital to understand the underlying signalling between the initiating factor, VEGF, and its eponymous receptor, VEGFR. Specifically, the primary receptor of VEGF-A is VEGFR2 [3, 5]. Primarily, this occurs through a canonical pathway whereby ligand-receptor interactions trigger receptor dimerisation and the auto-phosphorylation of tyrosine residues in the receptor intracellular domains. The newly formed phosphotyrosines thereafter serve as binding sites for adaptor molecules to commence a plethora of downstream signalling pathways, providing the origin of the term *cascade.* One interaction of notable importance is between PLCγ and ERK1/2, vital for the regulation of endothelial cell proliferation and migration during angiogenesis [11].

Notwithstanding the importance of PLCγ-ERK1/2, the primary regulation of this process is via modification of the quantity of active receptors and ligands. Given that a ligand and its complementary receptor have the capacity to interact if in close proximity, it is logical that varying the quantity of either component will modify the maximal biological response possible. In this case, VEGFR1 functions as a soluble decoy receptor to 'mop up' VEGF-A [12]. With no resulting signalling, VEGFR1 fundamentally diminishes the quantity of ligand available for interactions with its canonical signalling receptor, VEGFR2. Mechanisms of lesser importance include the regulation of VEGFR2 phosphorylation by tyrosine phosphatases, receptor endocytosis and trafficking [2, 7].

Conversely, co-receptors such as neuropilin 1 have been repeatedly identified as playing key roles in the enhancement of downstream signalling pathways of VEGFR2, e.g., p38 MAPK and ERK1/2 activation [13, 14]. Clinically, this implicates neuropilin 1 as an essential apparatus in neoplasm development, which relies on the expansion of an architecturally aberrant vascular network; and is evidenced by its elevated expression in a variety of human tumour biopsies, especially those that have metastasized [15].

18.2.3 Tip Cell Selection

To ensure organised and directional sprouting, one endothelial cell is 'selected' to lead the tip of the sprouting vessel, referred to as the *tip cell*, and will comprise the outgrowth of the new vessel towards the VEGF gradient [16]. Tip cells are characterised by their relative position in the vascular wall, modulation of cell-cell contacts, presence of filopodia, migratory behaviour and the ability to remodel the extracellular matrix [16].

During this process, the endothelial cells adjacent to the tip cells, known as *stalk cells*, adopt a proliferative phenotype and divide to elongate the sprouting vessel. Extensive cross talk between stalk and tip cells inhibits the selection of further tip cells in the local vicinity via a process called *lateral inhibition*, reaffirming the respective tip and stalk phenotypes [3, 6]. One example of this cross talk is Nrp1 signalling, which promotes tip cell formation through the expression of Smad2 and Smad3 genes, both of which suppress the stalk cell phenotype [17].

Importantly, it should be noted that this 'tip cell' profile is transient. Over a given period, a cell will dynamically switch between tip and stalk cell phenotypes according to changes in environmental cues [18]. Intercell competition for the tip cell profile occurs via the differential expression of VEGFR1 and VEGFR2 and subsequent upregulation of delta-like ligand 4 (Dll4).

A salient feature of a quiescent endothelium is a cell-cell equilibrium of Dll4-Notch signalling along the endothelial surface, whereby a single cell expresses both the Dll4 ligand and Notch receptors that are capable of interacting with Dll4 from adjacent cells [19, 20]. Endothelial cells utilise the binding of Dll4 to surface Notch receptors in order to negatively regulate angiogenesis via the downregulation of VEGF signalling [19, 21].

During angiogenesis, this equilibrium is disrupted. A local elevation of VEGF upregulates the surface expression of Dll4 on the newly selected tip cell. This initiates a tip cell-specific magnification of Dll4, increasing notch signalling to downregulate VEGF receptor expression in neighbouring cells [22]. This serves as the primary mechanism of inducing a stalk cell phenotype in the cells adjacent to the tip cell, limiting their migratory behaviour and retaining a connection to the existing endothelium, whilst also creating a precise sprouting pattern [22, 23].

Mechanistically, Notch receptor signalling facilitates the activation of ADAM γ-secretase in the tip-adjacent cells, catalysing the cleavage of the cytoplasmic Notch domain (Notch intracellular domain, NICD), which is now able to translocate to the nucleus and displace a repressor complex bound to the regulatory DNA region of the Notch target genes (e.g., hey, hes) [21]. The interaction of NICD with the DNA-binding RbpJ protein promotes upregulation of the Notch target gene expression, ultimately leading to an increased proliferation rate and the promotion of a stalk cell phenotype [21]. As a consequence of stalk cell Notch signalling, the expression of VEGFR2 is repressed while that of VEGFR1 is enhanced. As discussed, VEGFR1 exhibits a similar binding affinity for VEGF-A; however, it lacks additional signalling domains, instead functioning to 'mop up' VEGF-A in the vicinity of stalk cells, diminishing the amount of available ligand for VEGFR2 and thereby, decreasing Dll4 expression on stalk cells [12].

Notch signalling in the stalk cells also promotes Jagged-1 expression: an antagonistic ligand to Dll4. Jagged-1 is expressed on the stalk cell and interacts with tip cell Notch to prevent Dll4-Notch receptor signalling [24]. Any low levels of Notch still present on the surface membrane of tip cells can be bound to Jagged-1 in preference to Dll4, switching off the Notch signalling pathway to further establish a dominant tip cell phenotype [23, 24].

18.2.4   Sprout Extension and Anastomosis

Once the tip and stalk cell phenotypes are established, the sprout begins to elongate towards the gradient of VEGF in order to form a neovessel in the hypoxic tissue [16]. Cell migration is a complex process requiring the coordinated activity of the actin and microtubule cytoskeleton and the adhesion system. Under the tip cell phenotype, the migrating tip cell develops a polarised morphology, with filopodia and lamellipodia comprising an actin and microtubule cytoskeleton. These act in a coordinated fashion to establish cell motility and achieve directional migration based upon chemotactic gradients, namely VEGF [16]. Crucial to the regulation of cell migration are members of the small GTPase family of proteins including Rac1, RhoA and CDC42 [25].

Matrix metalloproteinases degrade the ECM, creating a physical path for the sprout to extend and migrate. This proteolytic remodelling produces a pro-angiogenic environment in which the sprout will extend until it meets another tip cell. Upon contact with another tip cell, adhesion junctions are formed by VE-cadherin, vital for the stabilisation of the newly formed vessel [10]. In a mechanism still poorly understood, resident tissue macrophages then promote the fusion of neighbouring tip cells, forming a physical connection between the two sprouts [26].

18.3   What We Don't Know: Vessel Patency

18.3.1   Sprouting Angiogenesis Lumen Formation

Once tip cells have fused, the sprout undergoes extensive remodelling to form a viable lumen for blood flow. Whilst this process is well established in vasculogenesis, the exact mechanism of lumen formation is yet to be fully elucidated in sprouting angiogenesis, with a number of competing theories. Historically, the *vacuole coalescence model* was widely accepted, whereby intracellular vacuoles coalesce, both with other vacuoles of the same cell and with vacuoles of neighbouring cells in order to create a single expanding intracellular vacuole that will eventually induce vessel patency [27, 28].

This was the general consensus until competing models arose in the mid-2000s. One such example is based upon the electrostatic repulsion of negatively charged glycoproteins on opposing apical endothelial cells.

This results in the de-adhesion of adjacent cells, creating a micro-slit as the earliest evidence of patency [29]. From this early lumen, VEGF-A mediates several changes in the architecture of apposing endothelial cells, increasing the lumen diameter through separation of the apical surfaces [29]. Finally, haemodynamic shear stress induced by the flow of blood through the primitive lumen further increases the cross-sectional area, creating a patent lumen. This theory, termed the *luminal repulsion by apical membrane model*, is supported by a strong animal-derived evidence base [30]. That said, newer theories have also been proposed and research is ongoing.

18.3.1.1   Reestablishment of Quiescence

The final stage of sprouting angiogenesis is the stabilisation of the newly formed vessel, characterised by the reattachment of pericytes, reformation of inter-cell junctions, and deposition of a new basement membrane.

The junctions between endothelial cells comprise claudins, junctional adhesion molecules (JAMs) and, most importantly, cadherins [10]. VE-cadherin is constitutively expressed at junctions where it mediates adhesion between adjacent endothelial cells and plays a vital signalling role. In quiescent endothelial monolayers, VE-cadherin clusters at junctions in zipper-like structures and exhibits frequent binding activity with β-catenin, p120 and plakoglobin [10]. In a non-quiescent endothelium, the phosphorylation of VE-cadherin and β-catenin tyrosine residues reduces junctional strength, causing this complex to become partially disorganised with increased permeability, e.g., facilitating leukocyte transmigration [10].

Through binding to catenins, VE-cadherin promotes cellular survival pathways. Upon the binding of ligands to endothelial frizzled receptors, subsequent Wnt-mediated signalling facilitates the detachment of β-catenin from junctional VE-cadherin [31]. Importantly, the now-cytosolic β-catenin avoids proteolysis due to inhibition of its degradation complex by Wnt signalling, allowing β-catenin to translocate to the nucleus and initiate proproliferation and pro-survival gene expression changes [32]. If VE-cadherin is lost at junctions, an expected accumulation of β-catenin will be accompanied by a marked decline in gene expression changes due to the absence of Wnt signalling and thus the continued presence of β-catenin degradation complexes. Therefore, the expression of VE-cadherin is essential for junctional and thus vessel stabilisation [33].

Mural cells (also known as pericytes) have been found to play a major role in neovessel stabilisation, primarily via the ANG1-TIE2 pathway. Mural cells attach to endothelial cells and synthesise TIMP3 and angiopoietin-1 (ANG-1), the latter of which then binds to TIE2 receptors on the vessel surface to induce Dll4 signalling in the endothelium [34]. This establishes Notch signalling in neighbouring cells of the neovessel, which in this case serves two main functions. Firstly, it acts to reduce the expression of VEGFR2, suppressing further sprouting angiogenesis [34]. Secondly, it upregulates NRARP expression, promoting 'tightening' of VE-cadherin junctions via a Wnt signalling apparatus [35, 36]. The propagation of equipoised unidirectional Notch signalling along the endothelial surface is a salient feature of a stable, quiescent vessel [6]. The process of angiogenesis is complete and a new viable blood vessel with functional capacity has been formed.

> **Take-Home Message**
>
> - Angiogenesis is normally switched off. When switched on, it has significant physiological importance in embryonic vascular development, wound healing and menstruation.
> - Regulation of angiogenesis is achieved through a carefully controlled temporal and spatial balance of stimulating (e.g., VEGF) and inhibitory (e.g., high Notch signalling) factors.
> - Inappropriate activation of angiogenesis can contribute to pathological processes (e.g., cancer, diabetic retinopathy and atherosclerosis).

References

1. Semenza GL (2007) Vasculogenesis, angiogenesis, and arteriogenesis: mechanisms of blood vessel formation and remodeling. J Cell Biochem 102(4):840–847
2. Simons M, Gordon E, Claesson-Welsh L (2016) Mechanisms and regulation of endothelial VEGF receptor signalling. Nat Rev Mol Cell Biol 17(10):611–625
3. Hiratsuka S, Kataoka Y, Nakao K, Nakamura K, Morikawa S, Tanaka S et al (2005) Vascular endothelial growth factor A (VEGF-A) is involved in guidance of VEGF receptor-positive cells to the anterior portion of early embryos. Mol Cell Biol 25(1):355–363
4. Carmeliet P, Ferreira V, Breier G, Pollefeyt S, Kieckens L, Gertsenstein M et al (1996) Abnormal blood vessel development and lethality in embryos lacking a single VEGF allele. Nature 380(6573):435–439
5. Shibuya M, Yamaguchi S, Yamane A, Ikeda T, Tojo A, Matsushime H, Sato M (1990) Nucleotide sequence and expression of a novel human receptor-type tyrosine kinase gene (flt) closely related to the fms family. Oncogene 5(4):519–524
6. Potente M, Gerhardt H, Carmeliet P (2011) Basic and therapeutic aspects of angiogenesis. Cell 146(6):873–887
7. Schuermann A, Helker CSM, Herzog W (2014) Angiogenesis in zebrafish. Semin Cell Dev Biol 31: 106–114
8. Huang J, Zhao Q, Mooney SM, Lee FS (2002) Sequence determinants in hypoxia-inducible factor-1α for hydroxylation by the prolyl hydroxylases PHD1, PHD2, and PHD3. J Biol Chem 277(42):39792–39800
9. Kaelin WG (2002) Molecular basis of the VHL hereditary cancer syndrome. Nat Rev Cancer 2(9): 673–682
10. Harris ES, Nelson WJ (2010) VE-cadherin: at the front, center, and sides of endothelial cell organization and function. Curr Opin Cell Biol 22(5):651–658
11. Srinivasan R, Zabuawala T, Huang H, Zhang J, Gulati P, Fernandez S et al (2009) Erk1 and erk2 regulate endothelial cell proliferation and migration during mouse embryonic angiogenesis. PLoS One 4(12):e8283
12. Shibuya M (2006) Vascular endothelial growth factor receptor-1 (VEGFR-1/Flt-1): a dual regulator for angiogenesis. Angiogenesis 9(4):225–230
13. Kawamura H, Li X, Goishi K, Van Meeteren LA, Jakobsson L, Cébe-Suarez S et al (2008) Neuropilin-1 in regulation of VEGF-induced activation of p38MAPK and endothelial cell organization. Blood 112(9):3638–3649
14. Lanahan A, Zhang X, Fantin A, Zhuang Z, Rivera-Molina F, Speichinger K et al (2013) The neuropilin 1 cytoplasmic domain is required for VEGF-A-dependent arteriogenesis. Dev Cell 25(2):156–168
15. Tse BWC, Volpert M, Ratther E, Stylianou N, Nouri M, McGowan K et al (2017) Neuropilin-1 is upregulated in the adaptive response of prostate tumors to androgen-targeted therapies and is prognostic of metastatic progression and patient mortality. Oncogene 36(24):3417–3427

16. Gerhardt H, Golding M, Fruttiger M, Ruhrberg C, Lundkvist A, Abramsson A et al (2003) VEGF guides angiogenic sprouting utilizing endothelial tip cell filopodia. J Cell Biol 161(6):1163–1177
17. Aspalter IM, Gordon E, Dubrac A, Ragab A, Narloch J, Vizán P et al (2015) Alk1 and Alk5 inhibition by Nrp1 controls vascular sprouting downstream of Notch. Nat Commun 6:7264
18. Jakobsson L, Franco CA, Bentley K, Collins RT, Ponsioen B, Aspalter IM et al (2010) Endothelial cells dynamically compete for the tip cell position during angiogenic sprouting. Nat Cell Biol 12(10): 943–953
19. Segarra M, Williams CK, De La Luz Sierra M, Bernardo M, McCormick PJ, Maric D et al (2008) Dll4 activation of Notch signaling reduces tumor vascularity and inhibits tumor growth. Blood 112(5): 1904–1911
20. Lobov IB, Renard RA, Papadopoulos N, Gale NW, Thurston G, Yancopoulos GD et al (2007) Delta-like ligand 4 (Dll4) is induced by VEGF as a negative regulator of angiogenic sprouting. Proc Natl Acad Sci 104(9):3219–3224
21. Blanco R, Gerhardt H (2013) VEGF and Notch in tip and stalk cell selection. Cold Spring Harb Perspect Med 3(1):a006569
22. Tammela T, Zarkada G, Wallgard E, Murtomäki A, Suchting S, Wirzenius M et al (2008) Blocking VEGFR-3 suppresses angiogenic sprouting and vascular network formation. Nature 454(7204): 656–660
23. Suchting S, Freitas C, le Noble F, Benedito R, Breant C, Duarte A et al (2007) The Notch ligand Delta-like 4 negatively regulates endothelial tip cell formation and vessel branching. Proc Natl Acad Sci 104(9):3225–3230
24. Benedito R, Roca C, Sörensen I, Adams S, Gossler A, Fruttiger M et al (2009) The notch ligands Dll4 and Jagged1 have opposing effects on angiogenesis. Cell 137(6):1124–1135
25. Raftopoulou M, Hall A (2004) Cell migration: Rho GTPases lead the way. Dev Biol 265(1):23–32
26. Fantin A, Vieira JM, Gestri G, Denti L, Schwarz Q, Prykhozhij S et al (2010) Tissue macrophages act as cellular chaperones for vascular anastomosis downstream of VEGF-mediated endothelial tip cell induction. Blood 116(5):829–840
27. Kamei M, Brian Saunders W, Bayless KJ, Dye L, Davis GE, Weinstein BM (2006) Endothelial tubes assemble from intracellular vacuoles in vivo. Nature 442(7101):453–456
28. Davis GE, Camarillo CW (1996) An α2β1 integrin-dependent pinocytic mechanism involving intracellular vacuole formation and coalescence regulates capillary lumen and tube formation in three-dimensional collagen matrix. Exp Cell Res 224(1):39–51
29. Strilić B, Kučera T, Eglinger J, Hughes MR, McNagny KM, Tsukita S et al (2009) The molecular basis of vascular lumen formation in the developing mouse aorta. Dev Cell 17(4):505–515
30. Ribatti D, Crivellato E (2012) "Sprouting angiogenesis", a reappraisal. Dev Biol 372(2):157–165
31. Klaus A, Birchmeier W (2008) Wnt signalling and its impact on development and cancer. Nat Rev Cancer 8(5):387–398
32. Olsen JJ, Pohl S öther G, Deshmukh A, Visweswaran M, Ward NC, Arfuso F et al (2017) The role of Wnt signalling in angiogenesis. Clin Biochem Rev 38(3):131–142
33. Yap AS, Niessen CM, Gumbiner BM (1998) The juxtamembrane region of the cadherin cytoplasmic tail supports lateral clustering, adhesive strengthening, and interaction with p120(ctn). J Cell Biol 141(3):779–789
34. Zhang J, Fukuhara S, Sako K, Takenouchi T, Kitani H, Kume T et al (2011) Angiopoietin-1/Tie2 signal augments basal notch signal controlling vascular quiescence by inducing delta-like 4 expression through AKT-mediated activation of β-catenin. J Biol Chem 286(10):8055–8066
35. Phng LK, Potente M, Leslie JD, Babbage J, Nyqvist D, Lobov I et al (2009) Nrarp coordinates endothelial Notch and Wnt signaling to control vessel density in angiogenesis. Dev Cell 16(1):70–82
36. Geudens I, Gerhardt H (2011) Coordinating cell behaviour during blood vessel formation. Development 138(21):4569–4583

18

Endothelial Function in Normal and Diseased Vessels

Mridul Rana, Zarius Ferozepurwalla, and Justin Mason

© Springer Nature Switzerland AG 2019
C. Terracciano, S. Guymer (eds.), *Heart of the Matter*, Learning Materials in Biosciences,
https://doi.org/10.1007/978-3-030-24219-0_19

What You Will Learn in This Chapter

This chapter provides a comprehensive overview of the endothelium, its physiology and its dysfunction. We will begin by exploring the origin of endothelial cells and their importance in the regulation of blood pressure, inflammation and vascular homeostasis. From there, leukocyte transmigration will be expanded upon, with a discussion on relevant novel research. Next, the interplay of haemodynamic factors, namely differential flow patterns, on the endothelium will be explored, including several key mechanosignalling facets. Finally, the role of nitric oxide and endothelin-1 in regulating vascular tone will be examined.

Learning Objectives

- Outline the origin of endothelial cells and describe the key differences between their activated and natural quiescent states.
- Provide an overview of the main pathways and molecules involved in leukocyte transmigration.
- Outline the concept of shear stress, its role in endothelial homeostasis and the response of key mediators (nitric oxide and endothelin-1) in regulating vascular tone.

19.1   Origin of Endothelial Cells

Establishment of the vasculature is critical for embryonic development and postnatal life, allowing for the delivery of oxygen and nutrients to cells, in addition to the removal of waste products. Lining the vasculature are endothelial cells (ECs), which arise from meso-dermal precursor cells (*haemangioblasts*) in the extraembryonic yolk sac and the embryo itself to form a de novo network of vessels in a process termed *vasculogenesis* [1]. Haemangioblasts form clusters known as blood islands, from which the outer layer gives rise to ECs. In their mature form, the endothelium comprises a single layer of flat cells that line the entire vascular and lymphatic systems [2]. These cells differ depending on their position within the cardiovascular system (CVS), with distinct transcription factor (TF) profiles allowing for the development of organ-specific ECs [1]. Broadly, the endothelium can be classified into three groups [2]:

- *Continuous*: ECs are tightly connected, with a continuous basement membrane. This is typical of the skin, muscle, lung and central nervous system (CNS).
- *Fenestrated*: the basement membrane contains areas of thinning where holes or fenestrae occur, classical of exocrine glands and the kidney.
- *Discontinuous*: marked gaps are present, with a poorly-formed underlying basement membrane allowing for high permeability. Found in the liver and bone marrow.

While different EC markers exist for different tissues, there is no universal marker present in all (and exclusively in) endothelial cells.

19.1.1   Resting Endothelium

The resting endothelium is quiescent and balanced towards an anti-inflammatory, anti-thrombotic, anti-apoptotic and anti-proliferative phenotype [3]. Activation of the endo-thelium disrupts this equilibrium in favour of a pro-inflammatory and pro-thrombotic state, the promoters of which include cytokines, low shear stress and smoking [3].

19

With ECs lining the entire circulation system, they play a significant role in cardiovascular health. Its aberrant, dysregulated activation, termed *endothelium dysfunction*, represents a critical step in the development of several cardiovascular diseases (CVD) including coronary artery disease, hypertension (HTN), stroke, peripheral vascular disease and venous thrombosis [3]. Collectively, these constitute a considerable global morbidity and mortality burden [3].

19.2 Endothelial Regulation

The endothelium plays a vital role in several aspects of physiology including angiogenesis, the local inflammatory response and haemostasis. Accordingly, dysfunction of this system is noted in the disease processes of several coagulation disorders (▶ Chap. 16) and almost all stages of atherosclerosis (▶ Chap. 17), to name a few.

19.2.1 Leukocyte Transmigration

Leukocyte transmigration arises during the local inflammatory response to trauma, infection, etc. and also pathophysiologically in atherosclerosis [4]. Leukocytes are recruited to the site to resolve the inflammatory stimulus and repair the surrounding tissue. Local chemokine releases from ECs and resident tissue macrophages, e.g., monocyte chemoattractant protein-1 (MCP-1), binds to receptors on inflammatory cells to promote their migration down a chemokine gradient [5]. Leukocytes undergo capture, rolling, firm adhesion and transmigration to enter tissues. These stages are summarised as follows.

19.2.1.1 Capture

Cell signalling proteins (cytokines) play a central regulatory role in inflammation, with two key mediators being tumour necrosis factor alpha (TNF-α) and interleukin-1 (IL-1). The presence of inflammatory cytokines upregulates the expression of P- and E-selectins by the endothelium, which bind to glycoproteins, e.g., P-selectin glycoprotein ligand-1 (PSGL-1) on leukocytes [6]. Of note, P-selectin is stored in EC granules and also in platelets, allowing immediate expression/release in response to pro-inflammatory stimuli [5]. This overexpression of selectins facilitates the formation of catch bonds, which capture leukocytes from the bloodstream, with the strength of the bond positively correlating with the force of blood [7]. This triggers a tyrosine kinase (TK)-mediated signalling cascade through the cytoplasmic tail, the outcome of which is an upright, high-affinity integrin conformation [5, 8]. Integrins are transmembrane proteins involved in intercell adhesion via their interactions with EC immunoglobulin ligands.

19.2.1.2 Rolling

The binding of PSGL-1 to selectins initiates a nexus of intracellular signalling pathways, e.g., the P38 MAPK pathway, which switch integrins into a high affinity state to facilitate their binding to ligands [9]. Two endothelial ligands of particular importance are [5, 10]:

- *Vascular Cell Adhesion Molecule-1 (VCAM-1)*: binds to 'very late antigen-4' (VLA-4) on monocytes.
- *Intercellular Adhesion Molecule-1 (ICAM-1)*: binds to 'lymphocyte function-associated antigen-1' (LFA-1) on monocytes.

Both of these adhesion molecules facilitate leukocyte rolling and are upregulated by inflammatory cytokines, especially TNF-α. There is stronger binding between leukocytes and the endothelium through this change in integrin conformation. This process is termed *inside-out signalling* [8].

19.2.1.3 Slow Rolling

Bond strength is then increased via *outside-in signalling*, whereby receptors cluster together to increase valency and the formation of stronger bonds with leukocytes via the recruitment of additional tyrosine kinase and PI3K (phosphoinositide 3-kinase) signalling [8, 11]. Moreover, outside-in signalling also regulates the subsequent steps of leukocyte transmigration.

19.2.1.4 Firm Adhesion

The eventual outcome of chemokine-mediated slow rolling is leukocyte arrest, assisted by a potpourri of leukocyte surface integrins [5]. A key interaction is between LFA-1 integrins and ICAM-3 on the endothelium, which results in Ca^{2+} mobilisation, adhesion and the secretion of further chemokines [12]. Finally, immunoglobulin-like domains on leukocyte surface ligands form strong disulphide (covalent) bonds with the endothelium.

19.2.1.5 Leukocyte Transmigration

The transmigration of leukocytes out of the vasculature relies on both 'docking structures' found on ECs that are rich in ICAM-1 and VCAM-1 [13], and a host of endothelial cell-cell junction molecules, three of which are [14, 15]:

- *Junctional Adhesion Molecules (JAMs)*: bind to integrins to propagate the movement of leukocytes through intercellular spaces. They are constitutively expressed at endothelial junctions and can be induced on the endothelial surface following inflammatory stimuli.
- *Platelet Endothelial Cell Adhesion Molecule-1(PECAM-1)*: interact with cytosolic catenins to facilitate migration.
- *Vascular Endothelial (VE)-Cadherin*: an adhesion molecule whose expression on opposing endothelial surfaces normally contributes to the formation of tight intercellular junctions via homophilic interactions [16]. This maintains the structure of the endothelium.

VE-cadherin is constitutively phosphorylated and then dephosphorylated. However, loss of phosphatase activity through leukocyte binding leads to VE-cadherin phosphorylation and resultant EC dissociation. This disruption of the endothelial integrity increases permeability to promote leukocyte transmigration [16].

Transmigration is either *paracellular* (in the intercellular space between cells) or *transcellular* (through a cell) [5]. Paracellular transmigration involves cells passing between cells without increasing permeability, possibly via the induction of Rho-GTPases by ICAMs, or by increased intracellular Ca^{2+} that binds to myosin light chain kinase to trigger EC contractions. In contrast, transcellular transmigration is thought to occur via the extension of leukocyte projections into ECs to trigger a complex cytoplasmic signalling system that facilitates vesicle organelle channels, through which the leukocyte can migrate. This entire process is summarised in ◘ Fig. 19.1.

19

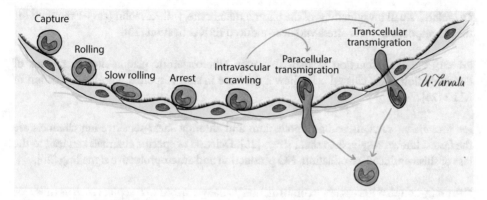

Fig. 19.1 Diagram summarising the canonical five-stage process of leukocyte transmigration, namely capture, rolling, slow rolling, adhesion and transmigration. Key interactions are shown, including VCAM-1/VLA-4 and ICAM-1/LFA-1

19.3 Mechanotransduction and Mechanosignalling

Shear stress is the frictional force per unit area that endothelial cells are constantly exposed to by blood flow [17]. *Linear laminar pulsatile flow* refers to the rapid, high shear stress profile of flow that runs parallel to the endothelium and activates several cytoprotective pathways.

The resulting signalling is anti-thrombotic, anti-adhesive, anti-proliferative, anti-inflammatory and anti-apoptotic and induces the synthesis of nitric oxide (NO), prostacyclin and tissue plasminogen activator (tPA) [18, 19]. Typically, soluble molecules released by endothelial cells in response to laminar flow help in maintaining a stable quiescent vessel wall. However, exceedingly high shear stress can injure and erode the endothelium, causing it to become prothrombotic [20]. In contrast, low shear stress occurs at the inner curvatures of vessels, whereby bifurcations and branch points cause disturbed blood flow, slowing its velocity and creating non-linear and turbulent patterns [18]. This oscillatory flow is pro-inflammatory and associated with plaque formation [19].

The entire process of mechanosensing and mechanotransduction in endothelial cells is termed *mechanosignalling* and can result in either the maintenance of vascular homeostasis or contribute to endothelial cell dysfunction and atherosclerosis.

19.3.1 Mechanosensors

ECs detect patterns of flow, and thus variation in shear stress, using a multitude of receptor types expressed on their surface. Let us explore some of the identified mechanosensors capable of detecting alterations in mechanical input/flow:

Glycocalyx   a glycolipid and glycoprotein complex that coats the inner endothelial surface. Alteration or removal of one of its constituents, heparan sulphate, impairs flow-mediated responses, including the induction of flow-mediated genes [21]. One heparan sulphate proteoglycan, syndecan-4, has previously been characterised as a critical mechanosensor whose loss is associated with suppression of KLF2/4 signalling and an increased susceptibility to atherosclerosis [21, 22].

Caveolae   small invaginations of the plasma membrane, with caveolin (cav)-1 required for the EC response to shear stress via shear-induced ERK activation [23].

Primary Cilia   dysfunction is associated with atherosclerotic plaque sites in regions of disturbed flow. Non-ciliated cells show a decrease in flow response, with less induction of KLF2 [24].

Ion Channels   calcium, sodium, potassium and chloride shear-sensitive ion channels are the fastest known response to shear stress [25]. Deletion of specific channels can lead to the loss of shear-induced vasodilation, NO production and atheroprotective signalling [26].

G Protein-Coupled Receptors   conformational changes occur to receptors such as those for bradykinin when exposed to shear stress [27]. Bradykinin agonists inhibit shear-induced mechanisms.

EC Nuclei   the nuclei of endothelial cells have been shown detect both the strength and direction of blood flow, ultimately leading to transcriptional and phenotypic changes in the EC [28].

Integrins   shear stress induces the rapid activation of select integrins. Blocking integrin signalling prevents the activation of shear-sensitive signalling pathways [29].

Junctional Proteins   PECAM-1, VE-cadherin and VEGFR2 (vascular endothelial growth factor receptor 2) are all required to form a shear-responsive mechanosensory complex [30].

19.3.2 Mechanosignalling

Following induction by mechanosensors, several pathways comprise the mechanosignalling response. Some of these include:

Krüppel-Like Factor 2 (KLF2)   the first transcription factor identified to be shear sensitive. The interactions between KLF2 and KLF4 play a role in mediating the anti-inflammatory, atheroprotective effects of laminar flow [31].

Senescence   disturbance of flow encourages senescence (growth arrest) of endothelial cells via the p53-p21 pathway. This can be inhibited through the histone deacetylase enzyme sirtuin 1, which is activated by restoration of normal flow patterns; it is also activated pharmacologically [32]. Interestingly, resveratrol in some foods and wine also promotes sirtuin 1 expression, reducing the senescence of ECs and thus protecting them [32].

CD59   inhibits the terminal complement pathway by preventing the formation of membrane attack complexes (MAC). Kinderlerer et al. observed upregulated CD59 expression in the presence of laminar shear flow, yet this was dependent upon other key regulators, e.g., KLF2 expression [33]. Disturbed flow did not show any significant induction of CD59 expression. Hence, laminar shear stress can protect against complement-mediated injury by increasing CD59 levels [33].

19.4 What We Don't Know: Sirtuin 3 in Mechanosignalling

Sirtuin 3 (SIRT3) is a mitochondrial enzyme with several key roles specific to cardioprotection, sprouting angiogenesis, EC metabolism and EC-cardiomyocyte interactions [34]. It is thought that in some circumstance, endothelial cells in the coronary vasculature primarily utilise glycolytic metabolism to save oxygen for the adjacent cardiomyocytes located in the heart [34]; Interestingly however, He et al. identified SIRT3-deficient ECs as exhibiting decreased rates of glycolysis and increased mitochondrial oxygen consumption [35]. This may limit available oxygen for cardiomyocytes, potentially resulting in cardiomyocyte dysfunction and necrosis.

Overexpression of SIRT3 improves insulin resistance in ECs, whereas downregulation, using small interfering RNA (siRNA), proved to increase mitochondrial ROS and reduce the insulin response in human umbilical vein ECs [36]. Endothelial-dependent vasorelaxation also diminishes SIRT3 deficiency that is found in obese human and mice subjects [36].

19.5 Vascular Tone

We have already discussed in previous chapters how the interplay of several different systems collectively regulates vascular tone, and the instrumental role the endothelium has. Accordingly, the importance of nitric oxide (NO) as a vasoactive regulator has been explored, and how its downregulated expression is a salient feature of early endothelial dysfunction. A thorough understanding of NO is vital to appreciate the critical role it has in modulating vascular tone.

19.5.1 Nitric Oxide

NO is synthesised by endothelial nitric oxide synthase (eNOS) normally bound to caveolin proteins located in cell membrane invaginations called *caveolae* [37]. Increased intracellular Ca^{2+} (of both extracellular and endoplasmic reticulum origin) increases the probability of Ca^{2+} binding to calmodulin. This causes a structural change in calmodulin, facilitating its binding to eNOS, causing eNOS to simultaneously detach from caveolin and activate [37]. In the presence of specific cofactors, e.g., tetrahydrobiopterin, activated eNOS then catalyses the synthesis of NO from L-arginine [37].

NO produces protective anti-thrombotic, anti-adhesive and anti-proliferative effects on the vasculature. Because of this, the dysregulation of NO represents an essential stage of endothelial dysfunction [38]. Briefly, this occurs when eNOS switches from generating NO to producing superoxides and hydrogen peroxide, a process termed *eNOS uncoupling* [38]. eNOS uncoupling occurs in areas where there is local excess of reactive oxygen species (ROS) that leads to the uncoupling of tetrahydrobiopterin from eNOS, with subsequent diminishment of nitric oxide biosynthesis [39].

Physiologically, NO diffuses from the endothelium into VSMCs, where it binds to and activates guanylyl cyclase, catalysing the conversion of GTP to cyclic GMP (cGMP) [37]. This acts to decrease smooth muscle tension via a reduction in Ca^{2+} release from the SR. This pathway has been targeted therapeutically, with the development of nitroglycerin (prodrug) and sildenafil (a phosphodiesterase-5 inhibitor that prevents cGMP degradation). Of note, laminar shear stress increases AKT-mediated phosphorylation of eNOS, enhancing its catalytic activity [40].

19.5.2 Endothelin-1

Another important modulator of vascular tone is endothelin-1 (ET-1): a peptide synthesised and secreted by ECs, which mediates some of its effects on these cells directly [41]. It contributes to both vascular tone and the regulation of cell proliferation through its binding to ETA and ETB receptors, present on cells of the CVS, namely cardiomyocytes, fibroblasts, smooth muscle and ECs, and also tubular and glomerular cells in the kidney [41]. Though previously considered purely as a vasoconstrictor, the current consensus is that ET-1 has an effect on vascular remodelling, angiogenesis and extracellular matrix synthesis.

- ET-1 acts at endothelial ETB receptors, inducing the production of NO and prostacyclin to mediate vasodilation and angiogenesis [41].
- ET-1 acts at ETA receptors on VSMCs to bring about vasoconstriction and proliferation. This is through Ca^{2+} acting as a second messenger, resulting in SR Ca^{2+} channel opening and subsequent vasoconstriction [41].
- ET-1 signalling activates macrophages, elevating the production of free radicals [41].

19.6 Clinical Implications: Replacement of Damaged Endothelium

Endothelial dysfunction and damage result in the loss of endothelial integrity. Consequently, endothelial cells detach and enter the circulation, allowing their utilisation as a marker of damage [42]. Replacement of the damaged endothelium occurs by either division of the surrounding ECs or by the actions of endothelial progenitor cells.

Cardiovascular risk factors, e.g., smoking, impairs the mobilisation of these endothelial progenitor cells, as well as their differentiation and function [43]. However, it has been demonstrated that increased numbers of these circulating ECs can offset the impairment resulting from cardiovascular risk factors [43]. Furthermore, exercise and statins have been shown to have a beneficial effect on the mobilisation of these progenitor cells [44, 45].

> **Take-Home Message**
>
> - Endothelial cells line the entire vasculature in a quiescent state unless activated to a pro-inflammatory, prothrombotic state. This change in phenotype contributes to a range of cardiovascular diseases that impose a significant health burden.
> - Leukocyte transmigration mediates their recruitment to areas of local inflammation to either aid defence systems or as part of a wider inflammatory pathophysiology.
> - Laminar/high shear flow exerts a cytoprotective effect on the endothelium, whilst disturbed/low shear flow is pro-inflammatory.
> - Endothelial dysfunction is characterised by an imbalance between vasodilatory and vasoconstrictive molecules including nitric oxide, prostacyclin, bradykinin and endothelin.

19

References

1. Schatteman GC, Awad O (2004) Hemangioblasts, angioblasts, and adult endothelial cell progenitors. Anatomical record – part a discoveries in molecular, cellular, and evolutionary biology. Anat Rec Part A 276A:13–21
2. McCarron JG, Lee MD, Wilson C (2017) The endothelium solves problems that endothelial cells do not know exist. Trends Pharmacol Sci 38:322
3. Rajendran P, Rengarajan T, Thangavel J, Nishigaki Y, Sakthisekaran D, Sethi G et al (2013) The vascular endothelium and human diseases. Int J Biol Sci 9:1057
4. Huo Y, Schober A, Forlow SB, Smith DF, Hyman MC, Jung S et al (2003) Circulating activated platelets exacerbate atherosclerosis in mice deficient in apolipoprotein E. Nat Med 9:61
5. Ley K, Laudanna C, Cybulsky MI, Nourshargh S (2007) Getting to the site of inflammation: the leuko-cyte adhesion cascade updated. Nat Rev Immunol 7:678
6. McEver RP, Cummings RD (1997) Role of PSGL-1 binding to selectins in leukocyte recruitment. J Clin Investig 100:485
7. Marshall BT, Long M, Piper JW, Yago T, McEver RP, Zhu C (2003) Direct observation of catch bonds involving cell-adhesion molecules. Nature 423:190
8. Kinashi T (2005) Intracellular signalling controlling integrin activation in lymphocytes. Nat Rev Immunol 5:546
9. Ellies LG, Tsuboi S, Petryniak B, Lowe JB, Fukuda M, Marth JD (1998) Core 2 oligosaccharide biosynthe-sis distinguishes between selectin ligands essential for leukocyte homing and inflammation. Immunity 9:881
10. Huo Y, Hafezi-Moghadam A, Ley K (2000) Role of vascular cell adhesion molecule-1 and fibronectin connecting segment-1 in monocyte rolling and adhesion on early atherosclerotic lesions. Circ Res 87:153
11. Giagulli C, Ottoboni L, Caveggion E, Rossi B, Lowell C, Constantin G et al (2006) The Src family kinases Hck and Fgr are dispensable for inside-out, chemoattractant-induced signaling regulating beta 2 integrin affinity and valency in neutrophils, but are required for beta 2 integrin-mediated outside-in signaling involved in sustained adhesion. J Immunol 177(1):604–611
12. Campanero MR, Del Pozo MA, Arroyo AG, Sánchez-Mateos P, Hernández-Caselles T, Craig A et al (1993) ICAM-3 interacts with LFA-1 and regulates the LFA-1/ICAM-1 cell adhesion pathway. J Cell Biol 123:1007
13. Barreiro O, Yáñez-Mó M, Serrador JM, Montoya MC, Vicente-Manzanares M, Tejedor R et al (2002) Dynamic interaction of VCAM-1 and ICAM-1 with moesin and ezrin in a novel endothelial docking structure for adherent leukocytes. J Cell Biol 157:1233
14. Muller WA (2003) Leukocyte-endothelial-cell interactions in leukocyte transmigration and the inflam-matory response. Trends Immunol 24:326
15. Nourshargh S (2006) The role of JAM-A and PECAM-1 in modulating leukocyte infiltration in inflamed and ischemic tissues. J Leukoc Biol 80:714
16. Dejana E (2004) Endothelial cell-cell junctions: happy together. Nat Rev Mol Cell Biol 5:261
17. Ku DN, Giddens DP, Zarins CK, Glagov S (1985) Pulsatile flow and atherosclerosis in the human carotid bifurcation. Positive correlation between plaque location and low oscillating shear stress. Arterioscler Thromb Vasc Biol 5(3):293–302
18. Cheng C, Tempel D, Van Haperen R, Van Der Baan A, Grosveld F, Daemen MJAP et al (2006) Atherosclerotic lesion size and vulnerability are determined by patterns of fluid shear stress. Circulation 113:2744
19. Pan S (2009) Molecular mechanisms responsible for the atheroprotective effects of laminar shear stress. Antioxid Redox Signal 11:1669
20. Casa LDC, Deaton DH, Ku DN (2015) Role of high shear rate in thrombosis. J Vasc Surg 61:1068
21. Florian JA (2003) Heparan sulfate proteoglycan is a mechanosensor on endothelial cells. Circ Res 93:e136
22. Baeyens N, Mulligan-Kehoe MJ, Corti F, Simon DD, Ross TD, Rhodes JM et al (2014) Syndecan 4 is required for endothelial alignment in flow and atheroprotective signaling. Proc Natl Acad Sci 111:17308
23. Park H, Go YM, Darji R, Choi JW, Lisanti MP, Maland MC et al (2000) Caveolin-1 regulates shear stress-dependent activation of extracellular signal-regulated kinase. Am J Physiol Heart Circ Physiol 278:H1285

24. Hierck BP, Van Der Heiden K, Alkemade FE, Van De Pas S, Van Thienen JV, Groenendijk BCW et al (2008) Primary cilia sensitize endothelial cells for fluid shear stress. Dev Dyn 237:725

25. Brakemeier S, Kersten A, Eichler I, Grgic I, Zakrzewicz A, Hopp H et al (2003) Shear stress-induced up-regulation of the intermediate-conductance Ca2+−activated K+channel in human endothelium. Cardiovasc Res 60(3):488–496

26. Boycott HE, Barbier CSM, Eichel CA, Costa KD, Martins RP, Louault F et al (2013) Shear stress triggers insertion of voltage-gated potassium channels from intracellular compartments in atrial myocytes. Proc Natl Acad Sci 110:E3955

27. Chachisvilis M, Zhang Y-L, Frangos JA (2006) G protein-coupled receptors sense fluid shear stress in endothelial cells. Proc Natl Acad Sci 103:15463

28. Tkachenko E, Gutierrez E, Saikin SK, Fogelstrand P, Kim C, Groisman A et al (2013) The nucleus of endothelial cell as a sensor of blood flow direction. Biol Open 2:1007

29. Shyy JYJ, Chien S (1997) Role of integrins in cellular responses to mechanical stress and adhesion. Curr Opin Cell Biol 9:707

30. Tzima E, Irani-Tehrani M, Kiosses WB, Dejana E, Schultz DA, Engelhardt B et al (2005) A mechanosensory complex that mediates the endothelial cell response to fluid shear stress. Nature 437:426

31. Nayak L, Lin Z, Jain MK (2011) "Go with the flow": how Krüppel-like factor 2 regulates the vasoprotective effects of shear stress. Antioxid Redox Signal

32. Warboys CM, De Luca A, Amini N, Luong L, Duckles H, Hsiao S et al (2014) Disturbed flow promotes endothelial senescence via a p53-dependent pathway. Arterioscler Thromb Vasc Biol 34:985

33. Kinderlerer AR, Ali F, Johns M, Lidington EA, Leung V, Boyle JJ et al (2008) KLF2-dependent, shear stress-induced expression of CD59: a novel cytoprotective mechanism against complement-mediated injury in the vasculature. J Biol Chem 283:14636

34. Halestrap AP, Clarke SJ, Khaliulin I (2007) The role of mitochondria in protection of the heart by preconditioning. Biochim Biophys Acta Bioenerg 1767:1007

35. He X, Zeng H, Chen ST, Roman RJ, Aschner JL, Didion S et al (2017) Endothelial specific SIRT3 deletion impairs glycolysis and angiogenesis and causes diastolic dysfunction. J Mol Cell Cardiol 112:104

36. Yang W, Nagasawa K, Münch C, Xu Y, Satterstrom K, Jeong S et al (2016) Mitochondrial sirtuin network reveals dynamic SIRT3-dependent deacetylation in response to membrane depolarization. Cell 167:985

37. Förstermann U, Sessa WC (2012) Nitric oxide synthases: regulation and function. Eur Heart J 33:829

38. Siragusa M, Fleming I (2016) The eNOS signalosome and its link to endothelial dysfunction. Pflugers Archiv Eur J Physiol 468:1125

39. Förstermann U, Münzel T (2006) Endothelial nitric oxide synthase in vascular disease: from marvel to menace. Circulation 113:1708

40. Boo YC, Sorescu G, Boyd N, Shiojima I, Walsh K, Du J et al (2002) Shear stress stimulates phosphorylation of endothelial nitric-oxide synthase at Ser 1179 by Akt-independent mechanisms. Role of protein kinase. A J Biol Chem

41. Kowalczyk A, Kleniewska P, Goraca A (2015) The role of endothelin-1 and endothelin receptor antagonists in inflammatory response and sepsis. Arch Immunol Ther Exp 63:41

42. Erdbruegger U, Haubitz M, Woywodt A (2006) Circulating endothelial cells: a novel marker of endothelial damage. Clin Chim Acta 373:17

43. Hill JM, Zalos G, Halcox JPJ, Schenke WH, Waclawiw MA, Quyyumi AA et al (2003) Circulating endothelial progenitor cells, vascular function, and cardiovascular risk. N Engl J Med 348:593

44. Laufs U, Werner N, Link A, Endres M, Wassmann S, Jürgens K et al (2004) Physical training increases endothelial progenitor cells, inhibits neointima formation, and enhances angiogenesis. Circulation 109:220

45. Vasa M, Fichtlscherer S, Adler K, Aicher A, Martin H, Zeiher AM et al (2001) Increase in circulating endothelial progenitor cells by statin therapy in patients with stable coronary artery disease. Circulation

Printed in the United States
By Bookmasters